# The
# Virtual
# Corporation

PREVIOUS BOOKS BY THE AUTHORS

**William H. Davidow:**
Total Customer Service
Marketing High Technology

**Michael S. Malone:**
The Big Score
Going Public

Structuring and
Revitalizing the
Corporation
For the 21st Century

# The
# Virtual
# Corporation

## William H. Davidow
## Michael S. Malone

Edward Burlingame Books/HarperBusiness
HarperCollins*Publishers*

HarperCollins books may be purchased for educational, business, or sales promotional use. For information please write: Special Markets Department, Harper-Collins Publishers, Inc., 10 East 53rd Street, New York, NY 10022.

FIRST EDITION

*Designed by Alma Hochhauser Orenstein*

Library of Congress Cataloging-in-Publication Data

Davidow, William H.
    The virtual corporation: Customization and instantaneous response in manufacturing and service; lessons from the world's most advanced companies / William H. Davidow, Michael S. Malone. —1st ed.
        p.        cm.
    Includes index.
    ISBN 0-88730-593-8 (cloth)
    1. Technological innovations—Management—Case studies. 2. Manufactures—Technological innovations—Management—Case studies. 3. New products—Management—Case studies. 4. Service industries—Technological innovations—Management—Case studies.
I. Malone, Michael S. (Michael Shawn) II. Title.
HD45.D33   1992
658.5—0020                                                    92-52569

92 93 94 95 96 ❖/RRD 10 9 8 7 6 5 4 3 2 1

**TO SONJA**

*Whose love, support, and patience
make life and family wonderful.*

**TO CAROL**

*Conscience, partner, and Muse*

# Contents

# Acknowledgments

*The Virtual Corporation* has been made possible by the efforts of others. Many of the clues, insights, and source documents key to this book came from the brilliant scholars at BRIE (Berkeley Roundtable on the International Economy)—John Zysman, Michael Borrus, and Stephen Cohen—as well as BRIE's friends and foils, including Benjamin Coriat, John Locke, Charles Sable, and Earl Hall. We also thank Donald Kennedy of Stanford University, venture capitalist Steve Herrick of the Herrick Fund, and marketing consultant Russ Berg for their thoughtful comments and criticisms.

The companies of Silicon Valley and their managements provided technological and organizational insights. Jim Morgan, CEO of Applied Materials, has almost created a virtual corporation. He served as a sounding board for us on management topics. John East, president of Actel, is running a technologically virtual corporation—he and his firm helped provide the underlying vision of this book. Intel and its top executives, Gordon Moore and Andy Grove, have done so many things right in preparation for the new business revolution that their work served as a basis for this book.

On a personal level, Stewart Putney helped us make our views concise and our prose clear. Janey Young, who has lived through numerous books with her husband, cheerfully attacked another one with her new employer. Once again, Kathleen Chambers came through with her precise transcriptions. And Louise Bullock of Custom Information Searches Inc. of San Jose helped answer some last-minute questions.

Finally, we wish to thank our agent, Don Congdon, our editor, Ed Burlingame, his assistant Christa Weil, and HarperCollins publisher Bill Shinker for their enthusiastic support of this attempt to glimpse the future.

**1**

························································

# A New Kind of Business

I n 1781 William Murdock, an assistant to instrument maker James Watt, developed a gearing system called a "sun and planet" that converted the piston motion of Watt's celebrated new steam engine into rotational power to drive a shaft. The impact of this simple power transfer scheme was, by any measure, astonishing. Wrote technology historian James Burke, "One year after the 1781 patent on the sun and planet system, every graph on the British economy begins a sharp upward curve."[1]

One reason for this immediate effect was that ironmaster Henry Cort used the Watt engine in 1782 in a new process for producing wrought iron. Cort's system, fifteen times more productive than conventional techniques, suddenly made it cheaper to build many structures with iron than with wood—setting off an explosion in the construction of buildings, machinery, bridges, and, in time, railroads.

The result of this and other applications of the new power technology was the Industrial Revolution, the most fundamental reorganization of mankind since the rise of agriculture and cities 7,500 years earlier. The revolution rent the fabric of society, then rewove it in an entirely different pattern. The structures of everyday life changed forever—and with them the nature of families, governments, cities and farms, language, art, even our sense of time. In many important ways,

1

today we don't even think in the same way as those who lived before the Industrial Revolution.

The nature of iron making had not fundamentally changed for several millennia, until Mr. Cort managed to improve production by a little more than one order of magnitude. Yet that fifteen-times improvement was enough to precipitate the most sweeping revolution in recorded human history.

The most important power source of our own generation is information processing. Semiconductor integrated circuits and computers, the steam engines of information processing, have, over the past thirty years, managed to match Cort's performance increase more than once each decade. The warehouse-size computer of 1945 can now be found in your digital watch—on a silicon chip the size of a baby's fingernail. Denos C. Gazis of the IBM Research Center, reflecting on the progress of processing technology, has estimated that "my own personal computer has about five times as much power and storage capability as IBM Research provided to our entire Yorktown laboratory when I joined the company in 1961."[2] In the time it takes to read this page, a modern supercomputer can perform one trillion multiplications.

This electrifying pace of change has no precedent in human memory. To speak of total efficiency improvements in the billions is literally beyond human imagination. Yet this is the power source at the heart of modern life. Such dizzying change cannot help but have a fundamental, permanent effect upon the world's industries and the people who work in them. As a result, we are in the midst of a new business revolution, one as sweeping and far-reaching as its predecessor. It is already being experienced in the offices of corporations throughout the world.

The goal of this book is to examine these transformations to see how they interrelate with one another toward a larger purpose—and in the process put forth a vision of the corporation of the twenty-first century.

The importance of this vision cannot be overstated. By the year 2015 the United States will either be a leader in this new business revolution or it will be a postindustrial version of a developing country. Either a nation of independent knowledge workers or a colony of economic serfs. It will either enjoy a high standard of living or suffer increasing impoverishment—it will be either an economy transformed or a graveyard of industrial skeletons.

We are not alone. This same scenario holds true for all economically advanced nations. The winners will catch and ride this great historic wave and pull away at an astonishing rate. The losers will be inundated.

## The Age of the Virtual Product

The centerpiece of this business revolution is a new kind of product. This product (or service), though it has roots in the distant, artisan past, can only be built now thanks to the latest innovations in information processing, organizational dynamics, and manufacturing systems. Most important, it can be made available at any time, in any place, and in any variety.

Some of these products already exist in our daily lives. For example, prescription eyeglass lenses are ground and placed in custom frames within sixty minutes by companies like Lenscrafters and Pearle Vision Express. Polaroid gave us sixty-second photography years ago, but even it has been largely surpassed by one-hour developing and printing of high-quality conventional photographs. Electronic cameras play pictures on a TV set a moment after they have been taken or print them in a few seconds at our desks. Camcorders create instant movies. Personal computers and laser printers have made instant desktop publishing a reality in millions of offices and homes around the world. Some of the world's most sophisticated products, electronic gate arrays, which used to take months to be produced in $100 million factories, can now be designed and produced on an engineer's desk in minutes on systems costing less than $10,000.

Instantaneous services are available, too. We can get the oil changed in our cars in ten minutes. Travel reservations are made with electronic speed. We can obtain cash instantly at ATMs and transfer millions of dollars over telephone lines in microseconds. The thought of waiting days for international checks to clear, a fact of life just a few years ago, is now all but inconceivable. Federal Express carries documents and packages across the country for us in hours, yet even its speedy document transportation service has begun to look slow with the advent of the fax machine. And Taco Bell's new Express fast-food stores aim at fulfilling each customer's order in less than twenty seconds.[3]

3

What these products and services have in common is that they deliver instant customer gratification in a cost-effective way. They frequently can be produced in diverse locations and offered in a great number of models or formats. That such instantaneous products and services are already becoming pervasive, a regular part of our daily lives, in no way diminishes their importance. On the contrary, it underscores their effectiveness. They are so remarkable, so distinct from anything that came before that they deserve a special name. In this book they are referred to as *virtual products*. The ideal virtual product or service is one that is produced instantaneously and customized in response to customer demand.

A virtual product (the term will be used to mean both physical products and services) mostly exists even before it is produced. Its concept, design, and manufacture are stored in the minds of cooperating teams, in computers, and in flexible production lines.[4] While the *perfect* virtual product can never exist, there is little doubt that many will come close. For example, the Japanese are striving to "virtualize" the production of automobiles by putting in place systems that will produce cars to domestic order in just seventy-two hours.[5]

A note on definition: Traditionally, for something to be virtual meant that it possessed the powers or capabilities of something else. In the late 1950s, scientists developed what they called "virtual computers"—machines quick enough to handle several users sequentially while giving each user the impression of being the only one using the computer. This added the connotations of interaction and adaptability to the term. Virtual computers seemed to the user to exist anytime and anyplace they were needed, which in time led to the phrase "virtual reality."

We have extended the idea of virtuality one step further to reflect what is going on around us today—formerly well-defined structures beginning to lose their edges, seemingly permanent things starting to continuously change, and products and services adapting to match our desires. Not only will virtual products have great value for the customer, but the ability to make them will determine the successful corporations of the next century.

That ability won't come easy, however. Building a virtual product will require a company to utterly revise itself, control ever more sophisticated types of information, and master new organizational and production skills. Doing that will change most contemporary firms so completely that what will emerge from the process, a "virtual corporation," will have little in common with what existed before. Again, such a firm in its purest form will never exist. But the advantage will lie with those firms that best pursue such a goal. The closer a corporation gets to cost-effective instantaneous production of mass-customized goods and services, the more competitive and successful it will be.[6]

## A New Kind of Company

The virtual corporation began as a vision of futurists, became a possibility for business theorists, and is now an economic necessity for corporate executives. All of this has occurred in little more than a decade. This not only underscores the inevitability of this new business model but also hints at the speeded-up sense of time that will characterize it.

In 1980, when Alvin Toffler first talked of "prosumers" (consumers who produced what they consumed ) and "de-massified" production, it seemed more wishful thinking than vision.[7] Too many pieces were still missing. For example, few people in this country understood the lean production technique of Toyota's Taiichi Ohno, with its ability to combine the benefits of mass production with quick market response. Just as significant, much of the technology required to create virtual products had yet to be developed. Perhaps most important, the information processing power needed to make virtual products and services a reality would remain too expensive for years to come. In particular, the price per millions of instructions per second (mip) of computer power would have had to drop from $1 million on mainframe computers to about $100.

Since that time, however, the price per mip has plummeted. By early 1992 Hewlett-Packard, for example, had introduced a new $5,000 workstation offering 35 mips, or $143 per mip.[8] Thus, the $100 milestone is now just months away. As a result, the virtual corporation, once just a specter upon the horizon, is now hurtling toward us. Whether we are prepared for it or not, the virtual revolution has begun.

What will a virtual corporation look like? There is no single answer. To the outside observer, it will appear almost edgeless, with permeable and continuously changing interfaces between company, supplier, and customers. From inside the firm the view will be no less amorphous, with traditional offices, departments, and operating divisions constantly reforming according to need. Job responsibilities will regularly shift, as will lines of authority—even the very definition of employee will change, as some customers and suppliers begin to spend more time in the company than will some of the firm's own workers. Says manufacturing expert Earl Hall, "This change in the nature of 'product' will cause blurring of functions which are now understood to be manufacturing, design, delivery, finance, marketing—indeed, a new meaning of 'company.'" He adds:

> The complex product markets of the twenty-first century will demand the ability to quickly and globally deliver a high variety of customized products. These products will be differentiated not only by form and function, but also by the services provided with the product, including the ability for the customer to be involved in the design of the product. . . . A manufacturing company will not be an isolated facility of production, but rather a node in the complex network of suppliers, customers, engineering, and other "service" functions.[9]

The only way to give a customer a truly virtual product, one that adapts in real time to the customer's changing needs, is to maintain integrated and ever-changing data files on customers, products, and production and design methodologies. That means new and more sophisticated forms of market research and new product designs that enlist and empower the customer in the design and production process itself. For many firms, one of the primary tasks will be to develop systems and software that will enable the customer to take on design responsibilities heretofore reserved for the company.

It is thus better to talk of the virtual corporation in terms of patterns of information and relationships. Building virtual products will require taking a sophisticated information network that gathers data on markets and customer needs, combining it with the newest design methods and computer-integrated production processes, and then operating this system with an integrated network that includes not only highly skilled employees of the company but also suppliers, distributors, retailers, and even consumers.

It is not surprising then that profound changes are in store for both the company's distribution system and its internal organization as they evolve to become more customer driven and customer managed. On the upstream side of the firm, supplier networks will have to be integrated with those of customers often to the point where the customer will share its equipment, designs, trade secrets, and confidences with those suppliers. Obviously, suppliers will become very dependent upon their downstream customers; but by the same token the customers will be equally trapped by their suppliers. In the end, unlike its contemporary predecessors, the virtual corporation will appear less a discrete enterprise and more an ever-varying cluster of common activities in the midst of a vast fabric of relationships.

For many firms the challenge of all of this change will prove too great. For some employees, the experience will be more traumatic than that of the changes demanded by past industrial transformations— though the threat this time won't be regimentation, exploitation, or dehumanization but unpredictability, lack of comfortable structure, and, simply, too much responsibility. Executive careers spent building power and influence may turn out to be superfluous. Workers content to put in their hours, to do their work and go home, may suddenly find themselves saddled with responsibility and control they never desired. And companies content to maintain the status quo indefinitely may not only encounter change but be forced to endure continuous, unremitting, almost unendurable transmutation.

## Meaningful Restructuring

The challenge posed by this business revolution argues that corporations that expect to remain competitive must quickly achieve mastery of both information and relationships. Technology by itself, without commensurate changes in the rest of the corporation, will fail. By the same token, this process of corporate revision must be both rapid and complete. There is no intermediate step: the only way to build a virtual product is to revise R&D, manufacturing, marketing, sales, service, distribution, information systems, and even finance—all must metamorphose.

The employees of the virtual corporation must change as well. Virtual corporations will require large numbers of highly skilled, reliable, and educated workers—people who can understand and use the new forms of information, who can adapt to change, and who can work efficiently with others. This requires the ability not only to read, write, and perform simple arithmetic but to analyze and engineer. Virtual corporations will thrive only in an environment of teamwork, one in which employees, management, customers, suppliers, and government all work together to achieve common goals.

But change for its own sake will not be enough. Already some well-known corporations are in dangerous straits for having restructured, reorganized, and blindly installed expensive computer and robotic systems in the belief that they were some sort of magic bullets for productivity. What is needed instead is a meaningful corporate restructuring, one that first understands the forces at work driving the industry toward the creation of virtual products and then implements change using the right equipment, organizational schemes, and operating philosophies to achieve the necessary results.

A good example of the difference between meaningless and meaningful restructuring can be found in the contemporary stories of General Motors and Xerox. In late 1991 General Motors, the world's largest automobile manufacturer, announced that in the face of losses amounting to $1,500 for every car it built in North America it would lay off seventy-four thousand workers and close twenty-one plants over the following four years. By comparison, Xerox, a company that just a few years ago was suffering a business malaise comparable to GM's, instead of blaming its workers or foreign predations for its dwindling market share set about reinventing itself.

GM's problems were a long time coming. Employing nearly a half-million people, depending upon an unwieldy thirty thousand suppliers, ineffectively spending $77 billion on new plants and equipment despite running at only 80 percent of capacity, taking twice as many worker hours to build a car as Ford, and suffering from a staff bureaucracy that was in many places twice redundant, General Motors has come to symbolize the bloated inefficiency of a contemporary corporation trapped in the rules of the past.

The failure of GM's efforts to modernize led directly to the announcement of the cutbacks, in itself exemplifying how little GM understood its own difficulties. About the only competitive strength the company still retained rested in the loyalty of its workers, suppliers, and long-time customers. Thus, in one stroke, GM destroyed what little trust it had left—and trust, as we will show, is the defining feature of a virtual corporation. The announced restructuring, which did not identify the jobs that would be eliminated, could only build divisiveness. Until the middle of the decade, GM employees will live every day under a sword. Making matters worse, within hours after the announcement, industry observers and pundits were predicting these cutbacks to be only the beginning—hardly news to warm the hearts of loyal customers and suppliers. How, in this charged environment, as it fragments each of its constituencies, can General Motors ever hope to regain lost ground against other foreign and domestic automakers that are racing to pull together employees, managers, suppliers, and customers into virtual corporations?

By comparision, Xerox is putting the pieces together. Recent moves by Xerox have reduced design cycles, improved quality, empowered workers, significantly reduced the supplier base while improving its relationship with those that remained, shrunk the manufacturing cycle, made the company more responsive to customer needs, and dramatically reduced costs. Unlike GM, Xerox entered the 1990s with a vision of itself and a renewed sense of purpose. The ride will be bumpy, and there will no doubt be setbacks, but Xerox has made dramatic progress in regaining lost business and is well on its way to securing its future.

Why will one corporate restructuring be successful and the other not? Because Xerox comprehended the long-term goal of becoming a virtual corporation. GM, on the other hand, did not understand the future. Its actions appeared to be driven more from management's desire for self-preservation through cost cutting than from a sincere interest in market responsiveness, customer needs, or the company's own employees. Where GM's restructuring was aimed at improving on an obsolete management system, Xerox realized that the world had changed and that to survive it had to abandon the past and take bold steps forward.

Similar decisions will be made at thousands of corporations throughout the world in the next few years. It is not hyperbole to suggest that the success of national economies will depend upon how many of their domestic corporations are prepared to make the right choice. Companies that make the tough, and correct, decisions to become virtual corporations—that put the pieces together—will have the best chance of building a platform for sustained success. This is as true for the United States as it is for every other developed nation of the world. As Roger Nagel and Rick Dove of Lehigh University have written, "The standard of living Americans enjoy today is at risk unless a coordinated effort is made to enable U.S. industry to lead the transition to [this] new manufacturing system."

## The Past as Prologue

For an explanation of why the United States, long the world's industrial leader, finds itself in the current predicament, it is necessary to understand our industrial history. Hierarchical management systems were developed to provide control over the railroads. The hierarchy served as an information network within which data were gathered and summarized. Results were fed up the management chain so decisions could be made, and the decisions were then relayed back down the organization so they could be carried out. It was a management technique ideally suited to an era when distant communications were difficult and computers did not exist.[10]

It seems almost insulting to have to point out that much has changed in the past century or so. Much of what we think of today as modern business practice is merely a dressed-up version of the management styles of post–Civil War America. Computers can gather most information more accurately and cost-effectively than people, they can produce summaries with electronic speeds, and they can transmit the information to decision makers at the speed of light. Most interesting for our purposes is that, frequently, this information is so good and the analysis so precise that an executive decision is no longer required. A well-trained employee dealing directly with the situation can now make the decision faster and in a more responsive fashion than the remote manager miles away. Anyone restructuring a company that does not take this new employee empowerment into account is not dealing with the future but is merely streamlining the past.

The production systems used in many of our manufacturing and service companies are also based upon hundred-year-old technology. They are an extension of the concepts of Frederick Winslow Taylor, inventor of scientific management, and Henry Ford, perfecter of mass production. In the days of Taylor and Ford, the best way to produce goods cost-effectively was to build them in rigidly tooled, mass-production facilities. Those facilities operated best if work was systematically analyzed and reduced to its most predictable steps. This in turn led to a production system that manufactured one best product in one best way. Rigidity was the key to amortizing investment in single-purpose factories and effectively using deskilled workers in the production process. Change merely raised cost.

The advent of computers, new forms of organization, flexible automation, design automation, and numerous other technological advances has now made it possible for companies to be market responsive and cost-effective at the same time. Rigidity no longer works in advanced economies. Flexibility and responsiveness, once a threat to efficiency, are now keys to competitiveness.

As a result of these developments, by the beginning of the next century the corporation as we have known it for eighty years will have largely disappeared, its few survivors mostly huddled in dwindling market niches. Those in competitive markets that delay, in some cases even refuse, the process of becoming virtual corporations will be swept away, their remnants seized, reorganized properly, and absorbed by fast-moving modern competitors.

Unfortunately, in the United States the very success of the last important business transformation created certain well-entrenched myths that now must be abandoned if the nation is to move toward an economy based upon virtual products. One of these is the belief, promulgated since the Second World War, that manufacturing is not important. Some leading intellectuals have argued that the future of advanced economies lies in deindustrialization, in becoming service economies. From this viewpoint, manufacturing and production wither to unimportance in the bright light of brain work.[11]

Exacerbating the problem, standard textbook business histories only reinforce this myth. As we know, the United States was originally an agricultural society. As it became industrialized in the mid- to late-nineteenth century, farm employment fell from 80 percent of the working population to just 3 percent. Meanwhile, employment in industry experienced a commensurate increase. But, as researchers Stephen Cohen and John Zysman have eloquently shown in *Manufacturing*

*Matters,* while farm employment dropped, farm productivity grew at an even greater rate.[12]

The Industrial Revolution didn't replace farming, it transformed it. In the process, the United States became the most important agricultural producer in the world, far more vital than any of the current so-called agrarian societies. By that measure, the United States is more an agricultural country now than it was in Jefferson's day. This would suggest that what really happened was that the United States never stopped being an agricultural country—rather, the country became *both* an agricultural and an industrial economy.

A similar transition occurred in the middle of this century. While the United States became more and more productive industrially, the service industries that supported manufacturers began to grow as well. As Alfred Chandler has noted in *The Visible Hand,* a history of the management of American business from 1850 to 1930, the increasing complexity of manufacturing equipment and their surrounding bureaucracies demanded a growing army of people to maintain them.[13] That meant lawyers, accountants, insurance agents, land and sea transportation workers, construction workers, and postal workers. And in time it called for air transportation, telecommunications, software engineering, and on and on in the panoply that constitutes the structure of the modern state. Moreover, the wealth generated by the industrial sector created a population that had money to purchase services. It could buy meals in restaurants, go on vacation, and pay for health care and the college education of its children.

Meanwhile, just as in agriculture, industrial productivity increased and fewer people were required for the same task. Employment in services grew, partly because this sector was less susceptible to automation. Not surprisingly, by the 1970s, as employment in services grew to more than 50 percent of the work force, pundits began to proclaim that the United States had become a service economy.[14] In fact, America at its zenith became the world's leading agricultural, industrial, *and* service economy.

There is danger in believing that a large economy such as ours can exist without a manufacturing base. The competitive troubles of the last two decades are evidence of this. With legislators, educators, and market watchers emphasizing a new society built around the service industry, with the top business-school graduates gravitating toward the more glamorous, nonindustrial firms, production skills in the United States began to decline. More and more executives came to believe they could succeed by creating companies that were empty marketing shells—businesses that purchased products from low-cost offshore suppliers and resold them into the domestic market.

With the loss of production skills it became increasingly difficult to build competitive products in the United States. Return on investment in plant and equipment declined. In many cases the captains of industry accepted defeat passively, as if this industrial decline were preordained. Michael Dertouzos, Richard Lester, and Robert Solow have commented on one of the best-known victims in this tragic cycle: "The history of consumer electronics is a history of successive retreats by American firms, with the result that foreign manufacturers have won an entire market without ever having to fight a pitched battle. The American companies may never recover the lost ground." They conclude:

> The loss of the consumer-electronics industry will have wider consequences. Autos and consumer electronics are places where companies learn mass manufacturing. They are, for example the major customers for robots; without consumer electronics it is more difficult to have a successful robot industry. Consumer electronics also buys nearly half of all the semiconductors sold in Japan, a market not available to American producers.[15]

Those authors point out that in industry after industry, as the United States has conceded the manufacture of products, the industries directly supporting those products have withered as well. With the loss of consumer electronics, the suppliers of components to that industry declined. As the consumer electronics and components industries decayed in the United States, the suppliers of capital equipment lost their home markets and the customers that would help them design the next generation of equipment. Entire industry segments were lost—a scenario repeated with varying severity in the automobile, steel, textile, and numerous other industries.

Elsewhere in the world, in nations rebuilding their industrial strength, scarcity and competition were forcing companies to take greater risks and experiment with new ideas in order to survive. In

Japan, for example, new types of factories, with their faster and more adaptable production processes, forced an efflorescence of innovative ideas about manufacturing. This resulted in a collection of striking new procedures, including radical readaptations of proven ideas, such as *kanban;* systematization of enduring social values, like *kaizen;* and the adoption of useful foreign ideas, such as total quality control. In Germany, young, midsize firms struggled to survive against giant international competitors and in the process became the world's leading niche marketers—while the few domestic giants remained competitive by finding a cooperative arrangement with unions. The industries of northern Italy, faced with limited resources combined with complicated social structures and strident labor, developed innovative customer–supplier networks and found a high-value competitive edge in ingenious industrial design.

American confusion and complacence in the face of international innovation can be seen in the figures on world trade. The U.S. trade balance declined from a surplus of $40 billion in 1976 to a staggering deficit of $130 billion in 1989.[16] In the process the United States shrank from being the world's biggest creditor to the world's largest debtor nation. This in turn had devastating secondary effects. For example, lost industries hurt government tax revenues and contributed to the country's budget deficit. The deficit in turn stressed the capital markets, drove up interest rates, and forced companies to look for higher rates of return. That helped drive more manufacturing out of the country. And as manufacturing left, more people began seeking jobs in the service industries, which generate fewer dollars of output per employee than does manufacturing.

One result of this shift, beginning as early as 1960, was the slowdown in the growth in output per employee in the United States until it consistently lagged that of Japan, West Germany, France, Italy, and the United Kingdom.[17] This alone would be troubling enough, but with it came a flat, if not declining, standard of living for most Americans between 1971 and 1990.[18] For the 1980s, according to Merrill Lynch economist Bruce Steinberg, the United States had "the worst productivity showing for any decade of the century."[19]

The reality of the decline in our manufacturing base has begun to sink in not only within academic institutions but with leaders in government and, most important, the citizenry. One need only look at the problems plaguing our mature industries in the Rust Belt, view the end of the economic miracle in Massachusetts, or watch the agony of mass layoffs within the computer and semiconductor industries to realize that the nation is confronted with a serious systemic problem.

Perhaps these historic missteps might be more bearable if only the myth of a service-only economy were true. Unfortunately for us, the same disturbing downward trends are now taking place within service industries as well. The terrible truth is that the service economy is not self-contained. In fact, it is profoundly dependent upon manufacturing—not only as a consumer of the offered services but as a generator of the wealth to purchase those services.

Thus, at the beginning of the 1990s, the cost of believing the service society myth has been a hollowing out of the U.S. economy toward a rickety structure decaying at its center.

## A Virtual Necessity

Given this deterioration, it can hardly be surprising that in the United States there is a creeping sense that there is something terribly wrong with our economy. The uneasiness comes from the intuition that the problem cannot be solved simply by another tax reform or shift in monetary policy. Working harder and longer does not appear to be the solution to individuals either unable to find work or already holding down more than one job. The economy itself seems in some way terribly disoriented and misdirected. We know that the commercial environment is evolving around us but we don't understand how it is happening.

If indeed what we consider modern business organization is obsolete, if many of our organizational and managerial concepts are a century old, then no matter how much we streamline the processes and dress the operation in modern jargon, the results can only end up short, our strategies anachronistic, even malignant.

In this new light it also becomes apparent that our government is not confronting the economic problems of the future. Rather, it is driving the country into the next century while still clinging to outdated beliefs of the last. It is protecting industries using obsolete methods of production rather than supporting those positioning themselves for the future. It is attempting to cope with a new world competitive order by rebuilding a society capable of dealing with the environment of the 1950s. That is one reason why much of the restructuring going on in business and most of the machinations in Washington are sadly superficial. They attempt to deal with the problem by confronting only the symptoms. Instead, the country must push, by creating an economy based on virtual corporations, toward a vision of economic success in the twenty-first century.

There isn't much time. The new business revolution is coming fast because there are no viable alternatives. It is emerging in the corporate trenches out of pragmatic necessity, the product of a collision between aggressive international competition and breathtaking advances in technology. Soon, advanced economies, because they depend heavily on exports and because technology is easily diffusible to developing countries, will be unable to compete in world markets using traditional methods of production. The only hope is differentiation through national infrastructure, a highly educated populace, and sophisticated information processing systems. Therefore, the economies of the United States, Europe, Japan, and elsewhere must proceed down the path to reshaping their methods of production. If they fail in their efforts, they will be deprived of the wealth needed to improve and maintain their standard of living. That erosion has already begun in the United States, where the middle class is shrinking and the underclass is growing.

When one accepts the importance of manufacturing, as well as its interdependence with service, for a healthy society the virtual corporation becomes the best hope for a strong, value-added manufacturing base to maintain—and even improve—our quality of life in an increasingly competitive world. Whether it is producing a product or a service (and they become increasingly synonymous in our information-based age) the virtual corporation alone has the speed and flexibility to cope with a business environment of custom mass-produced products and services and a pace of change that will make contemporary business life seem glacial.

A number of businesses in the United States are now recognizing this change and have joined—some with a few departments, others with the entire firm—the first wave of this new business revolution. Some have done so by adopting techniques of companies in other nations, others have depended upon homegrown innovations. Either way, in the process all have gained formidable advantages over their competitors. Their stories, the heart of this book, prove that adapting to this radical discontinuity in business history is possible. But they also offer a warning: those companies, organizations, institutions, governments, and nations that will not or cannot adapt to this change—and quickly—are already gravely threatened.

The virtual corporation, a matter of speculation just a few short years ago, has now become an economic necessity. That is why, to restore prosperity to our country, we must be the first to grasp this future. A virtual future.

## The Race to the Future

This book is a portrait of the emerging virtual corporation, the institution that will be central to the new business revolution—what it is, how it will function, and why it is important. It is a vision of the future that offers a message of hope, a path for restoring prosperity to our economy. In most ways, daily life will be better in this new environment, but the transition to the new institutions and the new ways of operating will be difficult for many, painful for others, and catastrophic for a few.

Much of what you will read in the pages to come will probably not be new. Scores of articles and books have been written about such topics as just-in-time supply, work teams, flexible manufacturing, reusable engineering, worker empowerment, organizational streamlining, computer-aided design, total quality, mass customization, and so on. But *The Virtual Corporation* will, we believe, for the first time tie all of these diverse innovations together into a single cohesive vision of the corporation in the twenty-first century.

We will show that these programs, devised independently over the course of the last few decades in different countries throughout the world, are in fact crucial, interlocking components of a larger structure, an organization that is greater than the sum of these remarkable parts. Furthermore, this book will show that the organization, the virtual corporation, that results from integrating these components is so extraordinarily adaptable and fast moving as to almost overnight leave traditionally organized competitors far behind.

In order to share this vision with the reader it will be necessary to review these innovative new programs as they have transformed different departments of the modern corporation, from the executive office to the research lab to the factory floor. It will also be necessary to look once again at the mind-boggling changes that are occurring to the engine of this new business revolution, information processing.

With each topic we will also show, through case studies, how virtual products already have affected companies across a wide spectrum of industries, some of them in quite unexpected fields. In addition we will discuss how these changes relate to the overarching vision of the virtual corporation.

In particular, this first portion of the book (chapters 2–4) will look in some detail at how industries have evolved to create virtual products and services. Some of these examples will be missing key ingredients because many businesses that have progressed down the path to becoming virtual corporations have yet to make all the changes that will be required of them in the future. This section will also examine the role that the information revolution is playing and will play in the development of the virtual corporation.

The second part of the book (chapters 5 and 6) will examine the processes that are making it possible to create virtual products and services. It will explore the impact of invention, review how virtual products are designed, and delve into the way they will be produced.

The third section (chapters 7–10) will deal with organizational issues. It will look at the new types of relationships that must exist among management, employees, customers, and suppliers. It will examine the redistribution of functions between these groups and the structural changes that must take place within organizations to make the virtual corporation possible.

The book will close (chapter 11) with a commentary on the changes that must be made in government policy and in the U.S. economy to foster an environment in which virtual corporations can prosper.

There is urgency in all of this. An ever-advancing standard of living has not been vouchsafed to any of the industrialized countries of the world. There is no manifest destiny, only a critical choice. If these countries are to maintain and improve their standards of living as well as solve their social problems, they must find a new dynamo to create wealth. The virtual corporation will be that engine.

Currently, many of the world's most brilliant economists are struggling to understand the industrial trajectories that advanced economies will follow as they battle to remain industrially competitive. The researchers at the Berkeley Roundtable on the International Economy (BRIE) are engaged in a study of successful industrial models throughout the world—the General Motors New United Motor Manufacturing Inc. (NUMMI) production process, the industrial districts in northern Italy, the lean manufacturing systems of Japan—in an attempt to discover the genetic code that drives success.[20] It is our feeling that underlying many of these systems are the elements of the virtual corporation and that success will most often occur where these industrial models have found a means to reconcile the techniques used to create virtual products with the unique character of their society.

When he learned of this book, Benjamin Coriat of the University of Paris explained the need to begin "thinking in reverse" in order to discover the essence of the virtual corporation. He stressed that companies would have to begin with the customer and then determine how the virtual corporation should be structured. In other words, one could only discover the future by backing into it.

Coriat was right. To become a reality the virtual corporation will require a different perspective, one that to our untrained eyes at times may seem illogical. Without a doubt, the new business revolution will be a shock to the system, a blow to our sensibilities. It will require new social contracts, ever-higher levels of general education, and a frightening degree of trust. But we have no choice. The virtual corporation stands before us, offering us our best chance for revitalizing the nation's economy and guaranteeing meaningful employment for our citizens. If we don't walk through its doors, our global competitors certainly will.

# 2

## An Emerging Idea

Business revolutions are rarely the result of the pull of new ideas alone. They also require a push from the fear of business annihilation.

One such impetus driving companies toward virtual products is profitability—or rather the life-threatening lack of it. As Glenn Haney, former CEO of the electronics research firm Dataquest, has noted, modern business has grown so competitive that any interesting new product is quickly beset by a host of competitors that rapidly turn that product into a commodity, drive down prices, and squeeze out profit margins. Any hope of regaining an edge through innovation is doomed, Haney adds, because "technology itself has become a commodity that crosses engineering disciplines and political boundaries with increasing ease."[1]

Combine this commoditization, Haney continues, with the growing cost of innovation, and suddenly manufacturers face a new reality that they might not see any return on their new product lines until the third or fourth product generations. Thus, they risk "fatal success," as he calls it—the tragedy of going broke with a wildly popular invention.

Faced with this paradox, the only way out, Haney believes, is for companies to bind customers to them for those three or four product generations needed to turn a profit. Increased responsiveness to customer needs will be one of the best ways to do this. For the consumer, it means quick delivery of the desired product. That is one reason why the Japanese are pursuing the seventy-two–hour car.[2] For industrial companies it means letting customers participate in the design of needed products and then quickly producing them to meet their needs. This is precisely the program upon which chip maker LSI Logic Corporation has embarked. In doing so, however, the company has had to travel some unanticipated paths.

Consider for example the rigid contractual procedures that used to govern LSI Logic's relationships with its customers. LSI discovered that it was capable of manufacturing and delivering new products to customers faster than those customers could move the contracts through their own legal departments. Says company chairman and CEO Wilf Corrigan: "Our most aggressive customers said to us, 'Look, stock the product on our verbal order because we can't process the paper to you and you can't process the paper internally fast enough for the time window we have to operate in.'"[3]

LSI Logic chose to forge ahead with production and let the paperwork follow weeks or months later. This policy certainly undermines the spirit in which most contemporary business contracts are written, that of protecting the parties against the failure or bad faith of the other. However, those matters will have long since been resolved by the time LSI Logic and its customer put ink to paper. So, America's leading producer of gate arrays now enters business relationships based upon trust, on the assumption that the other side is equally responsible and fair-minded. LSI Logic ensures that trust by developing long-term relationships built on a shared and interdependent future.

Needless to say, this is not the typical way that American companies do business, and many would be ill-advised to move away right now from their rigid and legally secure environments. But the LSI Logic example shows that firms are willing to take risks to engender customer and supplier loyalties and to achieve faster design and shorter time-to-market cycles. This important strategic move should not be lost on competitors doing business in traditional ways who believe they are protected by the old rules and ponderous legal documents.

## The Value of Time

A second impetus propelling the virtual revolution is the extraordinary acceleration of events taking place thanks to technology. We live in an era when even microseconds—millionths of seconds—have become too slow a measure of some computer operations. We talk now of nanoseconds—billionths of seconds.

Businesspeople have long appreciated the value of time. Customers are always frustrated with waiting; they want things yesterday. By the same token, every company wants to get its product or service to the market first. Alfred Chandler argues for the importance of such "economies of speed." He notes that, historically, "increases in productivity and decreases in unit cost (often identified with economies of scale) resulted far more from the increases in the volume and velocity of throughput than from a growth in the size of the factory or plant."[4] Hewlett-Packard has conducted studies demonstrating that, while an engineering cost overrun of 50 percent impacts overall profitability just 4 percent, a time delay of six months in project completion can result in a 32 percent loss in after-tax profit.[5]

The book *The Machine That Changed the World* provides startling data on the speed with which Japanese auto manufacturers develop new cars.[6] Typically, from design to first delivery, a car takes forty-six months and 1.7 million engineering hours in Japan versus sixty months and 3 million engineering hours in the United States and Europe. The Japanese have used this time advantage to design cars that more closely track the ever-changing desires of customers. In recent years they have become the world's leading innovators of new cars, one model of which, the Honda Accord, has become the best-selling car in the United States. Because car designs produced by the Japanese have such a short life span, on average the Japanese actually produce less of every model than either the Americans or the Europeans—five hundred thousand versus nearly two million.[7] Thus, by focusing upon time in their race to dominate markets, the Japanese car manufacturers have managed to be industry leaders and specialty suppliers at the same time.

An emphasis upon time has also been the key to success for Germany's thousands of midsize (*Mittelstand*) companies. Using computers and other forms of automation, these firms, despite their size, have become major global players by targeting high-profit niche markets and keeping competitors away with flexible production. "I can switch production to a different product in seconds," claimed one maker of motors and gearboxes for cranes.[8]

Perhaps the most comprehensive study of time-based competition has been done by George Stalk and Thomas Hout of the Boston Consulting Group.[9] While looking into firms (nonmanufacturing as well as manufacturing) that had given responsiveness the same value as cost and quality—that is, firms beginning to produce virtual products—Stalk and Hout discovered some startling statistics:

> During these investigations many closely held assumptions as to how costs and customers behave have been altered. Instead of costs going up as run-lengths are reduced, they decline. Instead of costs going up with greater investment in quality, they decrease. And finally, instead of costs going up with increasing variety and response time, they go down. Further, instead of customer demand being only marginally affected by expanded choice and better responsiveness, it is astoundingly sensitive to this better service—with the company that is able to set customers' expectations for choice and response very quickly dominating the most profitable segments of demand.[10]

The magnitude of this dominance is a clue to the impact virtual products will have. Stalk and Hout studied five companies—the Wal-Mart discount store chain, Atlas industrial doors, Ralph Wilson Plastics, Thomasville furniture, and the mortgage department of Citicorp. What they found was stunning:

- A responsiveness to customers that was one-third faster than that of their nearest competitors;
- A rate of growth at least three times as great as the industry as a whole; and
- Profitability at least twice, and up to five times, as great as the competition.

Meanwhile, despite this explosive growth, Wal-Mart, for example, was still able to maintain the same service levels with one-fourth the inventory investment—one reason being that it empowered individual stores to order directly from suppliers, even overseas, thus reducing restocking time from an industry average of six weeks to just thirty-six hours.[11] Atlas could fill an order for an out-of-stock door in three to four weeks, compared with the industry average of three to four months. Most remarkable of all, Citicorp Mortgage had reduced the time for processing a loan commitment from the industry average of thirty to sixty days down to fifteen days or less, and then to as quick as fifteen minutes in some cases.[12]

Given these figures, is it any wonder that these time-based corporations were leaving the competition behind? The time-based competitors, said Stalk and Hout, "are literally running circles around their slower competitors."

And what was the reaction of the slower competitors? "Bafflement," according to Stalk and Hout. Trapped in traditional organizations and strategies, these firms watched helplessly as their speedy competitors

- Increased customer dependence;
- Skimmed off the most attractive customers (that is, impatient customers willing to pay a premium for promptness);
- Set the pace of industry innovation;
- Grew faster with higher profits; and
- Left the competition to try to catch up in a less supportive business environment.[13]

During the new business revolution this time-centered scenario will play itself out in industry after industry, not always in the same way but always with the same result.

All of this suggests a vision of a prosperous future—it means that there *is* a way. We are not doomed to reexperience the failures of the recent past.

But to this one must add that the path to becoming a virtual corporation will be far more difficult than the old way of doing business. It will demand changes of both rules and heart. But it also offers a way to break out of the inertia, a Sargasso Sea, that seems to have engulfed our society and economy. The coming business revolution offers a new purpose, a new goal, and, most important, a way to regain the economic leadership in a fashion that is both exciting in its style and humane in its manner.

This new business revolution may be a necessity for our society's survival, but it is also an enthralling opportunity. After all, America's history, more than that of any other nation, has been one of setting off into unexplored territory in search of a better life. "Change," says Jim Morgan of Applied Materials, "is the medium of opportunity."[14]

A number of businesses already are heading down the path to become virtual corporations and in the process experiencing enormous changes. None have proceeded far enough to come to grips with all of the difficult organizational and relationship issues. But even now, their unfinished stories, a sampling of which are offered in the rest of this chapter, give a clue to the extraordinary effect of "going virtual."

## The Arms Race

It is a measure of the harsh nature of business that few even well-known firms survive more than a few decades. Those that have endured since the Renaissance period amount to no more than a handful. By reviewing the changes that have taken place in such a firm one can learn much about how business transformations occur over time.

It is fortunate to have, in the work of Professor Ramchandran Jaikumar of the Harvard Business School, the recorded history, spanning five hundred years, of Beretta, the Italian small arms manufacturer.[15] The types of transformations that have taken place at Beretta over the course of a half-millennium now must occur in just a decade for modern firms racing toward becoming virtual corporations.

Founded in 1492, Beretta has been run by the same family for fourteen generations. This stability, combined with a product that has changed little in five centuries (there is little fundamental difference between a harquebus of 1675 and 9-mm automatic of 1975), makes Beretta an almost ideal test case for the impact of new manufacturing technologies and management techniques on the history of corporate enterprise.

In his research, Jaikumar identified six epochs in the story of Beretta. Each took approximately ten years to assimilate, and it was during those few distinct decades that the company made most of its

25

gains in productivity and quality. Furthermore, Beretta invented none of these new programs but learned quickly from its competitors, so that, in Jaikumar's view, "all of the changes were triggered by technology developed outside the firm."

This is perhaps the most important point to be gleaned from the Beretta example. One of the great weaknesses of Western business practice is the not-invented-here attitude, an unwillingness to accept new ideas from beyond the company walls. By the same token, the willingness of Beretta and Japanese companies, especially, to do just that has been one of their greatest strengths. The competitive race will require the assimilation of a broad range of technologies and techniques invented and developed by others. In the future it will be important to search the developed world for new ideas and capitalize on them quickly, as Beretta has throughout its history.

The story of Beretta begins with almost three hundred years of unchanging manufacturing techniques. All the company's guns were made by hand, by master gun makers who typically worked without plans from a basic model using a caliper. Apprentices watched the process and learned. All the activities in the company revolved around the notion of fit: parts were made and modified to fit tightly with other parts. Every gun was unique, parts could not be interchanged among different guns.

Three centuries of complacent inertia were shattered by the rise of scientific engineering at the end of the eighteenth century. Since then, Beretta has raced to stay abreast of each new wave of industrialization, Jaikumar's six epochs.

1. *The English System (1800)*—The Industrial Revolution reached Beretta during the Napoleonic Wars. Until about 1800, Beretta had built its weapons by hand using various jigs, clamps, and files. But the English System, centered on the notion of tools, separated the production's function from the processes used to make it. The result was universal fabrication tools such as metal lathes. Apprentices now were taught proficiency on a particular tool rather than a particular product, which in turn, says Jaikumar, meant that "process took on a life of its own, enabling process improvements to be made independently of product constraints."

The English System marked the arrival, though still halting, of mass production. Workers were expected to have fewer universal skills and more that were specific and tool-centered. In time, this led to industry built upon fully interchangeable, deskilled workers.

[The lean manufacturing facilities of the virtual corporation will reverse that trend, as they will depend on skilled workers with intimate knowledge of the process—just as Beretta did before the nineteenth century.]

2. *The American System (1850)*—The Industrial Revolution in the United States brought with it the need for high-volume production of products with interchangeable parts. Fittingly for a firearms maker like Beretta, the story began with Eli Whitney's contract to produce rifles for the U.S. Army in 1798. Beretta itself was impelled to use the American System after seeing the success of the Colt factory in London. In 1860 the firm acquired a complete American-Style manufacturing system from Pratt & Whitney.

The primary impact of the American System was the mechanization of work, with interchangeability only a by-product. With it came product specialization. After 350 years of being able to produce a theoretically infinite array of products, under the American System Beretta was reduced to making just three models.

[By adopting the American System, Beretta also accepted a rigid production process that still characterizes most modern mass production. The efficiencies that resulted have served as a rationale for avoiding product customization in many industries.]

3. *Taylor Scientific Management (1900)*—Frederick Winslow Taylor set out to do with labor what had already been done with machine tools; that is, to make it efficient, specialized, and interchangeable. This was accomplished through the redesign of factory floors and machine tools according to time–motion studies of individual laborers. The new role of management was to scrutinize the equations produced by this worker–machine–process interaction and determine the most efficient organization to deal with it. Interestingly, this greater efficiency (and product shift) also allowed Beretta to increase its product offerings from three to ten.

Traditional job responsibilities were broken down and given to multiple specially trained workers, worker discretion was replaced by the one best way to perform a task, and management controlled all aspects of work, comparing it with a predetermined standard.

[Taylorism widened the gulf between labor and management and ingrained in management the notion that direct laborers had little to contribute to the running of the corporation; their role was to take directions quickly, precisely, and in silence. As such, Taylorism represented a new mind-set, a corporate ethos that has led to generations of labor–management confrontation, institutionalization of the status quo, and an increasingly unskilled work force. The residues of Taylorism, more than any other business philosophy, have to be eliminated before a company can successfully produce virtual products.]

4. *Statistical Process Control (1945)*—In the 1950s, Beretta was licensed by the newly created North Atlantic Treaty Organization (NATO) to manufacture the Springfield M-1 Garand rifle, the mainstay of the U.S. Army in World War II.

With that license came the demand that the Beretta M-1s not only achieve parts tolerances greater than the company had ever known, but also that the rifle parts foster perfect interchangeability. This meant that not only would Beretta have to build new manufacturing equipment for the task, but it would have to develop a means of regularly sampling the output of these machines to ensure they didn't deviate beyond proscribed limits. Thus, Beretta would have to implement some form of statistical management over its production output.

Statistical quality control radically altered the organization of work at Beretta. Accurate performance by the machines was assumed, and only deviations were scrutinized; unlike the Taylor model, there was no one best way to operate, only a series of new problems to be detected and solved. With the machine running on its own, the operator was free to manage long-term performance. Quality engineering had arrived. The automaton worker of the Taylor era was now working with others in problem-solving teams monitoring machine performance.

[The era of worker involvement so essential in virtual corporations had begun. Perhaps even more important, the notion of quality control was introduced. As we shall see later, it is impossible to produce virtual products without perfect quality.]

28

5. *Numerical Control (1976)*—This epoch marked the arrival of information processing to the arms industry, first by computers and then by microprocessors.

Numerical control machines, which could automatically perform in sequence tasks that previously had taken multiple pieces of equipment, completed the 150-year progression from manipulation of a physical product to the processing of pure information.

Numerical control continued the retrograde labor movement away from Taylorism. Now, line workers were just half the total company employment. But their span of control was up to five times greater, encompassing several machines at once in what might be called a cellular organization of the manufacturing floor. That in turn demanded workers much better trained than at any time since the eighteenth century.

Beretta's 9-mm Parabellum won the biggest plum of all: the contract to replace the U.S. Army's Colt 45. Beretta won the contract because the new equipment allowed for a bid price half that of the nearest competitor and because the transportability of numerical control programs enabled the company to meet the army's stipulation of full U.S. production.

With numerical control, Beretta evolved from a user of information to an information-based corporation where much of the data required to manufacture products was stored in digital form on computers rather than on blueprints and in dies and molds.

[One of the keys to producing virtual products is the integration of production processes with suppliers producing perfectly interchangeable products. A good example of this occurs in the industrial districts of Italy where manufacturers have successfully distributed component making to scores of smaller contractors by making sure those suppliers use identical pieces of capital equipment that, when run with identical software programs and tooling, produce identical results. The introduction of numerical control enabled Beretta to export production processes in a similar fashion.]

6. *Computer-Integrated Manufacturing (1987)*—About the time of the army contract, Beretta was engaged in its most recent epochal transformation. This was the linking together of the entire

29

company within a computer network to perform computer-aided product design in engineering and a so-called flexible manufacturing system on the factory floor.

In practice, this meant that a company factory became a computer-controlled team of semi-independent workstations connected by automated material-handling systems of looped conveyors that carried pallets bearing the individual work pieces. A supervisory computer, carrying information about these work pieces, directed their progress through the manufacturing process, assigning priorities, queuing them, reacting to changing situations, assuring that all the necessary tools were on-line, loading the right numerical programs in the proper machines, and monitoring the resulting work as it occurred.

What has been the impact of computer-integrated manufacturing (CIM) at Beretta? For one thing, a three-to-one jump in productivity. The number of machines now required to produce a single product has now fallen to just thirty, the lowest in 150 years. The minimum number of people for efficiency is now just thirty as well, less than were needed at the end of the seventeenth century. Meanwhile, rework has been reduced to zero and staff positions represent two-thirds of Beretta's employment. Manufacturing is now treated as a service, customizing its products to the desires of special market segments. That in turn demands highly skilled "knowledge workers."

Within a few years, Beretta expects the process to be so entirely automated that, like some current Japanese factories, it will operate without human participation—in the dark, as it were.

What is perhaps most interesting about the current CIM epoch is that, for the first time since the guild days of the company three hundred years ago, Beretta is again theoretically capable of creating numerous different products. Customization and craftsmanship have returned. Beretta has come full circle.

## Regaining the Past

Another example of how a corporation can cost-effectively restore craftsmanship to manufacturing, and how technology can be put to the service of a long-established wisdom of the market, can be found at Beretta's competitor, Remington.

Like the account of Beretta, the history of Remington, though three and a half centuries shorter, is one of adapting to one industrial innovation after another.

Gun making at Remington, from the Civil War to the middle of the twentieth century, was a story of increasing, but limited, refinements on the mass production of machined interchangeable parts. Said one observer, "The steam engine was replaced by the dynamo and overhead belt drives by electric motors, but methods of making guns remained remarkably unchanged. If you could have visited the Colt or Remington or Winchester factories in 1870 and then returned in 1950, you would have been astonished at how little their methods of manufacture had changed."[16]

Remington knew it had to modernize to survive—and that meant finding some way to cut the cost of machining. One way the firm did so was to simplify its designs. Another was to use numerical tape-controlled machines, which it put in place on the manufacturing floor. These machines taught Remington about the power of automation, and so, in the mid-1980s, the firm jumped on the idea of computer-controlled machining—building a complete, hands-off factory for the job.

As one might expect, the results were stunning: "In the recent past, it took 24 machining operations to make a Remington shotgun receiver. That meant the part had to be manually positioned in separate holding fixtures and manually moved from cutting machine to cutting machine. Because of all of this cutting and shifting, it took six days to convert a block of steel into a finished receiver. . . . It now takes only about *four hours!*"[17] New models that used to take eighteen months to bring to production now take just six, and the manufacturing floor can be quickly shifted over for short runs of popular products.

So far, this is yet another positive story about the use of computers to improve productivity. But what makes the Remington story special is what the company did next.

A complaint of gun lovers is that modern rifles just aren't made like they used to be. The demands of modern manufacturing have resulted in simplified, Spartan designs. Receivers and other parts just didn't have the fit and feel of those in older, hand-machined and -fitted guns. An entire market of collectors was created to buy and trade these older weapons, the participants so precise as to demand items like pre-1964 Model 70 Winchesters.

Remington saw this market as an opportunity that, for the first time, could be tapped using its new computer-controlled machines. Said H. K. Boyle, Remington's plant manager, "We can reintroduce some of the old favorite models enjoyed by hunters and target shooters in the past, models that we stopped producing because they were too expensive to manufacture on standard machinery equipment."

Remington began this program of mining its own past by making original Parker shotguns for the first time in decades. It has proven to be a success. Other classic models are in the works. As pleased gun lovers have noted, these "new old" weapons aren't reproductions but continuations of the original production. As illustrated by this example, the processes used to produce virtual products need not only be a way of building the new and revolutionary products of the future, they can also be a way of bringing back the past in response to customer interest.

## From Monastery to Desktop

Bringing virtual products to the printing industry has taken more than a half-millennium, from monks carefully coloring the last medieval illustrated texts to modern color laser printing.

The famous transition point in printing from the laborious hand-made text to the low-cost mass production of the printed word occurred, of course, with the Gutenberg movable type press of 1440. It is important to note that this technological breakthrough was itself the product of several other technological breakthroughs—notably, low-cost paper and ink production and (introduced by Gutenberg himself) reusable molds for the type elements.[18]

The next five hundred years saw refinements of the Gutenberg process, such as the arrival of mechanical power and of new alloys that made high-speed, hot-type presses possible. This in turn created a revolution in information distribution, as the economies of print made it possible to produce books, magazines, and newspapers affordable to a growing percentage of the population.

The onrush of technology began to force a trade-off between efficiency and aesthetics. The advantages of the technology were obvious: the average person could now own a library greater than any Renaissance prince and could obtain in a single daily newspaper more information about the world than a Roman Caesar would have learned from a year's worth of dispatches. The words of great writers and statesmen could now reach out to more than just a favored few.

Lost in the process, however, was a certain control of result. One immediate victim of industrialized printing was an element of beauty. No published book of the seventeenth, eighteenth, or nineteenth century could compete for beauty with hand-illustrated works like the *Trés Riches Heures* of the Duc de Berry or with the Book of Kells.

Not that these losses were overly missed in the explosion of print that began in the mid-1700s. The sheer volume of printed matter overwhelmed most such considerations. Furthermore, printing technology itself continued to advance, notably with the addition of black-and-white, then color, photographs with the printed text. High-speed printing also allowed for multiple daily editions of a single paper; and an efficient national postal system made possible the weekly newsmagazine.

The next advance came with the arrival of phototypesetting, when the first Photon machine (Lumitype in Europe) was introduced in 1954.[19] Phototypesetting was important because it began the shift of printing away from the purely physical process to an informational process. Now the typesetter could sit at, say, a Mergenthaler machine and type copy, choosing from scores of typefaces stored on sheets of film and then photographically print out the result for stripping into the layout. Thanks to this greater speed of production, lower costs, and a vastly larger inventory of type fonts, control and creativity began to return to the printing business.

The other great change, beginning in the 1930s and reaching the small print shop by the 1960s, was a shift from mechanical to photographic technologies. Now a single typesetting machine might store hundreds of different typefaces. The result was not only a slashing of the time from typeset to pasteup to print, but much greater flexibility in the use of type and imagery. Costs fell too, such that short runs of sophisticated printing became reasonably affordable to even the smallest clients. Print shops soon found that offering in-house graphic design was both a useful marketing tool and a profit generator.

As important as these twentieth-century inventions were to the profession of printing, they were minor compared with the arrival of computer technology—microprocessors—beginning in the mid-1970s. The effect of this new technology was profound and nearly complete, from the replacement of typewriters with word processors at the low end of the market to the actual destruction of print shops by turnkey systems at the top.

The arrival of computers also signaled the advent of nearly instantaneous printing. Suddenly, print was no longer constrained by physical blocks of type or even strips of film. It was now encoded into bits of data that could be easily stored, manipulated, and transmitted to diverse locations.

The most striking effect on the printing industry came with the digitalization of the typesetting machine and the graphics department. Like many such paradigm shifts, this one began on the periphery and, as if from nowhere, suddenly and unexpectedly overturned the status quo.

The shift began with the Apple Macintosh personal computer. The Mac was an attempt by Apple to stave off the juggernaut of giant IBM by playing off Apple's image of being "user friendly." The Macintosh was designed to be the most transparently operable computer of its time and would include the use of a crisp, bit-mapped display, a mouse-driven cursor, and, to spare the new user the need to memorize a vocabulary of keyboard commands, graphic command icons that could be manipulated directly on the screen.

The Mac lived up to its goals, becoming one of the most successful personal computers ever built. Consumers loved it, but even more important, software designers became intrigued with this new graphical user interface as a way to explore applications impractical with cryptic keyboard commands. Within a few years, third-party designers were offering elementary programs for mixing text with images directly on the Mac screen for printout on dot matrix and, as they became affordable, laser printers.

Who was suddenly doing all this desktop publishing? Laypeople. The average computer owner no longer had to contract publishing work to a printer, inspect layouts, edit bluelines, and then wait weeks for the printed results. Now, for many applications, the same steps took minutes. Certainly—and this is an important point about the production of virtual products by customers—desktop publishing demanded a greater participation and understanding by the user. But, as billions of dollars in sales proved, the typical user was willing to make that sacrifice in exchange for control, speed, and lower cost.

Needless to say, some of the traditional bread-and-butter revenue sources for many printing houses began to disappear. But the desktop publishing industry did far more than steal existing business. Even more important to the rise of this market was the proliferation of new applications that desktop publishing had made feasible, such as the subsequent explosion in corporate, institutional, and personal newsletters.

Technological transmutations, however, are rarely confined to a single market segment. As the printing industry was losing some of its market at the low end, new technologies were enhancing its appeal at the top. The same microprocessor driving a personal desktop publishing system also could be used to build powerful new typesetting and layout machines. Those systems could run sophisticated software from firms such as Scitex and Hell Graphics to produce high-volume, magazine-quality printing with much the same speed and facility of their PC counterparts. With these programs, images could be cropped, modified, color balanced, and adjusted in tone to correspond with their neighbors, and entire publication issues could be laid out on the computer screen, stored, and then transmitted for printing.

Where did this printing take place? Anywhere. By the late 1980s, it was not unusual for a publication to be prepared in, say, California, printed in Taiwan, and distributed in Europe. These new technologies made it possible as well to target mass-market magazines and newspapers for ever-smaller regions until they approached perfect user configurability. It is easy to see that process already beginning with such publications as *USA Today*.

Even this won't be the end of the transformation of the printing industry. Why print at all? With broadband telecommunications, image compression, full-motion video transmission and storage on CD-ROM and laser disk, and portable color liquid crystal displays, within a generation it probably will be cheaper to receive all this information from a hand-held terminal than it will from newsprint—information that is not only fully customized to the particular interests of each consumer but backed by libraries of supporting material for curious minds. Hints at what is to come can be found in Apple's promotion of just such a terminal, to be called the Knowledge Navigator, and Knight-Ridder's current investigations into videotex and nonprint newspapers.

The direction then for the printing industry is toward consumer choice. The consumer will be able to obtain information matched to his or her needs faster, and in the format desired. Newspapers and books will never actually go away—they are, after all, in Borges's words, "mankind's imagination"—but the format of many will soon change from paper and print the way they once did from Sumerian clay and Egyptian papyrus scrolls.

There are several insights we can draw from the history of printing. The first is that virtual products will result from combining numerous and diverse technology advances. For example, the new printing processes are dependent upon lasers, xerography, integrated circuits, the microprocessor, high-speed communication processes, display technology, and advances in software.

The role of the author as a coproducer is also evident. He or she does not merely create the content but also can control the presentation of what is printed. To do that the author must not only be skilled in language but must also be computer literate. This is the new trade-off: greater control demands a corresponding extension in skill.

## History Etched in Silicon

The semiconductor industry created the modern computer industry, then fed on its dynamics. Therefore, any story about the transformation of the semiconductor industry should begin with the history of computers. The interaction of the two is also a reminder that creating virtual products is usually the result of the interaction between multiple, often unrelated, technological advances.

Computation machines, notably the abacus, have been around since prehistory. But the world's first electronic computer, the legendary ENIAC of 1945, is the standard by which all subsequent computers are measured. It also exemplifies the physical obstacles to building the early computers.

The challenge lay in the most elementary component of all computers: the switch. In the binary code of computers, all information is stored as a 1 or a 0—on or off. Hence the usefulness of the switch. The trick is making those switches reliable, small, and fast enough to keep up with the demands of high-speed computation. To do this, the designers of ENIAC were prepared to use vacuum tubes, which were essentially electrical switches with no moving parts. Even then, as writer Dirk Hanson has described it in *The New Alchemists,* ENIAC was an unwieldy beast:

> The common legend, no doubt apocryphal, but nonetheless telling, is that the lights of Philadelphia dimmed when ENIAC was switched on. For ENIAC was truly an electronic monster, an engineer's nightmare that was more room than a machine. . . . ENIAC had eighteen thousand tubes, seventy thousand resistors, ten thousand capacitors, six thousand assorted switches, and a maze of connecting wires. It measured one hundred feet long, ten high and three feet deep, and weighed in at thirty tons. . . .
>
> Tube reliability proved to be every bit as difficult as the engineers had figured. They rigged all manner of fans and blowers to carry off the tremendous heat generated by eighteen thousand vacuum tubes crammed in behind the metal panels, and still the temperature in the ENIAC room soared to 120 degrees F.[20]

In computations for atomic research, meteorology, and ballistics, ENIAC more than proved the value of computers; but it also convinced designers that they needed a more efficient type of switch to move the technology much further. The answer was already being found at Bell Laboratories in New Jersey, where John Bardeen, Walter Brattain, and William Shockley in 1947 used the new science of semiconductors to build the first transistor.

By the mid-1950s, the transistor had begun to revolutionize electronics, including computation. It then fell to two scientists, Jack Kilby at Texas Instruments and former Shockley protégé Robert Noyce at Fairchild Semiconductor, to determine that not only could the transistor be lithographically printed and etched onto a flat surface, but this process could be miniaturized to produce ever-larger, "integrated" transistor arrays. The result was the integrated circuit, perhaps the most important invention of the twentieth century.[21]

In retrospect, one can see the integrated circuit as perhaps *the* major step in the transformation of the information industry. It was very small, extremely reliable, consumed almost no power, and could be produced in great quantity for almost no cost. Thus, it could not only bring new levels of price and performance to computers but reduce their size so they could fit almost anywhere without expensive power and cooling apparatus.

The early 1970s saw the introduction into the market, most notably at Noyce's new Intel Corporation, of the microprocessor, an integrated circuit that brought together on its surface a computer built from logic, memory, and communications functions previously found only on collections of discrete chips. Putting a computer on a chip enabled the semiconductor industry to bring a new level of convenience to its customers. Now the actual performance of the device could be customized via programming. Engineers could change electronic systems by merely rewriting their programs rather than redesigning the hardware.

So, by making at least a partial shift in dependency from hardware to software, the microprocessor increased customer participation and, in the process, strikingly reduced the time between need and fulfillment.

Still, even this wasn't enough, especially as increased competition in the computer and other electronics businesses was leading to ever-shorter product generations—and greater demand for new state-of-the-art chips. This pressure quickly exposed the primitive nature of the chip-making business.

In essence, until the early 1970s the semiconductor industry was more a chemical business than it was involved in electronics. The extraordinarily complex designs for new circuits were commonly drawn by hand like blueprints on wall-size sheets of paper—a process that often took months. Then, these drawings were converted to photographic masks by cutting their design into rubylith and then reducing the image for printing on a silicon wafer. In many ways, the rubylith mask designers were to the early semiconductor industry what monks were to medieval printing.

By the end of the 1970s, with the demand for semiconductor devices reaching hundreds of millions of units each year, it was obvious that such primitive design techniques were wholly inadequate—and the process began to evolve at every step.

The biggest bottleneck by far, however, lay in the design process. Here the problem was the most challenging: how do you enhance human capability to reduce the design of a million transistor arrays from decades to weeks, while retaining the underlying creativity that makes the device useful?

The answer is encompassed in computer-aided design (CAD) and computer-aided engineering (CAE). These technologies have not been as simple to implement as the names might suggest; it was not enough just to give an engineer a computer and expect overnight break-throughs. Rather, the implementation of technology in engineering took place in a series of steps, each relying upon the ones before it.

The first of these steps was to computerize the creation of the rubylith artwork. This was done first through digitizers and later, by companies such as Calma, through powerful (and expensive) computer layout systems that featured large displays, extensive disk-based mem-ory, and complex languages and programs to convert operator com-mands into layout geometries.

The wonder of the interaction of computers and the semiconduc-tor industry is that its effects were cumulative. Computers made it pos-sible to create ever more complex integrated circuits that could be used to build more capable and less expensive computer systems—which in turn accelerated the process and reduced the cost of engi-neering more complex and cost-effective integrated circuits.

The software progressed as well—the cryptic early languages were replaced by English-language-type instructions. Also, the soft-ware could constrain the operator from straying beyond the capabili-ties of the circuitry. Best of all, the newest workstations could actually simulate the operation of the circuit under design, thereby allowing for real-time modifications and obviating the need for multiple, expensive prototype production runs. Suddenly, getting a million transistor devices to work the first time went from an impossibility to an every-day occurrence—and then to a competitive necessity. (These method-ologies are probably more advanced in the semiconductor industry than in any other. The design of semiconductors is a harbinger of what is to come in many mechanical design processes used for automobiles, airplanes, and buildings.)

By the mid-1980s, so many product designs had been run on such CAD/CAE systems that it was possible to begin developing libraries of circuit designs. Now, when a designer went to develop a new device, he or she didn't have to start with the basic switch elements but could work with a handful of building blocks and mix and match them to best fit the application.

The impact of these new design methodologies and equipment was striking. Design cycles for new chips dropped from years to months, even though the newer products were often hundreds of times more complex than their predecessors.

But this was only the beginning. As competition heated up among users of semiconductors, these firms increasingly demanded proprietary products from the chip houses. Not only did they want these devices to have performance characteristics different from the off-the-shelf versions sold to their competitors, but they also often wanted to design these custom products quickly and with minimal expense.

Custom chips, however, were an expensive proposition and required great skill to design. Even with CAD/CAE and automated fabrication techniques, producing just a few prototypes might cost more than a million dollars even before volume production. In response to this need, new companies sprung up to provide application-specific integrated circuits (ASICs), based on new types of design methodologies called gate arrays and standard cells. The goal of these companies was to provide mass-customized products quickly in response to customers' demand. Silicon Valley's LSI Logic and VLSI Technology are two prime examples.

ASICs combined two trends. One was the need to find a middle ground between stock and custom circuits to save on design and prototype costs. The other was the rapidly declining cost of integrated circuits. The reasoning behind these new devices was straightforward: if silicon circuits were cheap, why not be a little sloppy in the efficient use of a chip's surface in exchange for a radical reduction in design cost and quicker time to market?

Much of the production of these new ASIC companies was based on gate array technology. The trick here was that companies found ways to build custom circuits using standardized lower layers on the chip surface. They then customized the last few layers to generate the specific devices the customer wanted. What this meant was that wafers could be preprocessed and then warehoused to await the final customizing steps. Since it required only a few more processing steps to finish the circuit, much of the processing time to customize the circuit was eliminated for the customer. Furthermore, the cost of customizing the circuit was greatly reduced because the customer had to pay only to tool the final few layers.

Suddenly the cycle time to produce a custom part dropped from four to five months to as little as two weeks. Meanwhile, the cost of tooling custom circuits fell from hundreds of thousands of dollars to under $20,000—so cheap that ASIC manufacturers frequently gave away the tooling costs in their rush to capture business.

Many of these new ASIC firms were also as customer responsive as any companies in the world. Customers not only defined the products they wanted, they designed them, using workstations armed with intuitive, high-level languages. Design cycles and simulations were becoming so efficient that some daring customers began to eschew test runs and ordered full-scale production from the start, in order to pare time-to-market to the minimum.

As the eighties ended, an even newer generation of chip companies began to appear on the scene. The latest advances in information technology now made it possible for firms like View Logic Systems to offer CAE software that could be run on top-end personal computers. That meant the designer could create a new device on his or her desk. A new generation of ASIC firms, including Xilinx and Actel, developed integrated circuits—field-programmable gate arrays (FPGAs)—to take advantage of this capability. By using FPGAs, an engineer could "write" ASICs at his desk the same way a fellow employee down the hall might write the corporate newsletter.

Thus, in just two decades, the creation of integrated circuits passed from teams of trained specialists working months plotting out new designs to be fabricated in $100 million laboratories to a virtual product that could be designed in hours and "built" in minutes by an engineer who never had to leave the office, using a desktop "factory" costing less than $10,000.

The story of the integrated circuit industry illustrates many points about virtual products. One is the role of invention: virtual products in this field would not have been possible without the invention of the gate array. Just as important has been the transformation of the design process and the role of the computer in creating tools. The process is highly automated, designs can be accurately simulated, tooling can be computer-checked to eliminate errors, much of the engineering is reusable, and the process has become so efficient as to consistently make extraordinarily complex devices work perfectly the first time they are built. The design process has become inextricably tied to production. Data is fed directly from the design process to the machines that fabricate the tools that build the circuits. In some recent cases even tools aren't required; IBM, for example, has used electron beams to directly write patterns on wafers the same way images are projected on television screens.

With the development of virtual products by the semiconductor business it is now possible to build electronic systems on a chip in a few weeks for a few thousand dollars. A small silicon chip can replace a room of electronics that would have taken years to build just four decades ago.

## Scheduling the Skies

Much of the focus of this book has been and will be on products. But virtual services are of equal importance. One obvious example involves our system of exchanging money. As has been much reported, we are moving toward a cashless society that depends upon an electronic monetary system. Money already races around the world at the speed of light. Each day billions of dollars change hands in informational transactions that require neither gold nor currency nor printed documents.

Another service industry to benefit has been travel. Traveling has never been a simple process. From ships, coaches, and ferries to bullet trains and supersonic aircraft, each new technological breakthrough has brought with it a host of service challenges, many of them in ticketing and reservations.

Perhaps the most extreme example of this has been in the airline industry. With scores of airline companies and thousands of planes in the air crowding major lanes and runways, traffic control and airport management have required ever more sophisticated computers, software, and communications networks to keep up.

In 1989, for example, in the United States alone 351,272,900 customers booked seats on 12,843 daily domestic flights of major U.S. airlines.[22] Just keeping track of that information—combined with layovers, connecting flights, and seating assignments—is difficult enough. Add to this special meals, multiple timed connections, frequent-flier miles, and myriad discounts. Without powerful computer software, dealing with these millions of permutations would be impossible. The story of how the ticketing system for airlines, at first a carryover from passenger train systems, evolved into a virtual service is a remarkable one.[23]

The story of airline ticketing begins in the 1920s with the comparatively simple task of filling the solitary spare seat on airmail planes. Ticketing was simply a matter of taking down the name of the passenger who had paid for the flight on a given day.

The rise of multipassenger airliners in the early 1930s made the task more complicated. Handling reservations required keeping track of available seats and recording passenger names. By the end of the decade, new young airlines such as American were developing ways to overcome reservation confusion and overbooking by centralizing control at a flight's point of departure. This led to the so-called Request and Reply systems, which required the customer to contact the ticketing agent at the departure point and then wait for a return confirmation call.

The end of World War II brought with it the modern age of air travel and an explosion in demand that outstripped the supply of available flights. As individual seats became more valuable, so did the importance of quickly returning cancellations to the available pool for sale. The immediate solution to this was to install availability display boards in reservation offices so that agents could quickly scan flights for seat openings or, if booked, seats on alternative flights. Soon, however, explosive demand outstripped even this technology, until life as an airline ticketing agent took on an element of sheer absurdity. This was a typical scene at the American Airlines Chicago office in the mid-forties:

> A large cross-hatched board dominates one wall, its spaces filled with cryptic notes. At rows of desks sit busy men and women who continually glance from thick reference books to the wall display while continuously talking on the telephone and filling out cards. One man sitting in the back of the room is using field glasses to examine a change that has just been made high on the display board. Clerks and messengers carrying cards and sheets of paper hurry from files to automatic machines. The chatter of teletype and sound of card sorting equipment fills the air.[24]

In 1945 American Airlines set out to revamp its entire ticketing and reservation system. The push was directed by American's president, C. R. Smith, who had seen the power of new technologies to manage aircraft during his wartime stint as deputy commander of the Military Air Transport Command. He announced, "We're going to make the best impression on the travelling public, and we're going to make a pile of extra dough just from being first."[25]

Smith's ally in the project was Charles Ammann, head of American's Advanced Process Research Department, who had spent the war determining the need for such a system to remove the onus from ticketing agents, maintain an accurate running inventory, reduce the need for chaotic reservations offices, and automatically advise all offices of flight sellouts.

In visiting data processing vendors, however, Ammann soon found that while all were adept at building solutions for accounting applications (for example, one person needing access to thousands of records), few had any experience—or interest—in inventory control (hundreds of people needing access to a single record). So Ammann set out to build a system of his own.[26] One of his early prototypes used metal tubes that held marbles, to represent flights holding seats. Each time a seat sold, an electrical trip would drop one marble. Another prototype used resistive networks in a crude electrical system.

After many prototypes, the system that Ammann finally presented to Smith used a matrix of relays (columns were dates, rows were flights) in which shorting plugs could be manually inserted to indicate a sellout. Called the Reservisor System and built by Teleregister Corporation, it was installed at American's Boston reservations office in February 1946 and represented the "first time any airline had adapted current electronic discoveries to reservations handling."[27]

Within a year, the Boston office was handling an additional two hundred passengers daily with twenty fewer agents. The remaining agents worked at electrical keypads in which they keyed in their requests and saw a light blink if it were accepted or a still-dark screen if not.

The Reservisor was a considerable improvement over previous systems, but it was still slowed by the time it took to physically transfer passenger name record (PNR) cards and insert the shorting plugs. Maintaining records remained a sizable problem.

The next refinement to the system was the Magnetronic Reservisor, installed by American at LaGuardia Airport in 1952. It used the new magnetic drum computer memory as a low-cost storage system.

By 1956, a more sophisticated version, capable of handling two thousand flights for thirty-one days and which cut response times to one half-second, was installed at American's New York West Side Terminal. For the time being, the reservation inventory obstacle was overcome.

Now Ammann and his group attacked the challenge of passenger information. The result, first tested in 1956, was the Reserwriter, a computer that read punched cards filled with passenger data, converted them to paper tape, and transmitted the data over teletype. By 1956, the Reserwriter systems were linked into a nationwide network.

Despite obvious performance improvements, American's reservation system was still error prone. An estimated 8 percent of all transactions were in error, and to process a single round-trip reservation between New York and Buffalo required twelve people, fifteen procedural steps, and three hours. Furthermore, throughout most of the 1950s, productivity per reservations employee fell nearly 40 percent. Misplaced PNR cards too often led to overbooked flights and angry customers.[28] American knew that the pending jet age, with its greater seating capacity and shorter time windows, would only make the situation worse.

The solution, though American didn't yet know it, was already in the works. IBM, looking for new markets for its primarily military computers, had been eyeing Teleregister's success in airline reservations for years. The code name for Big Blue's search for consumer applications for its computers, terminals, and disk drives was SABER, and it was as part of this effort that the company approached Ammann at American. In 1956 a combined task force defined the needs of a state-of-the-art reservations system, and beginning in mid-1958 IBM set out to build it.

By the end of 1959, American had accepted IBM's proposal—though not without some trepidations about an untried telecommunications system that would cost $16,000 per agent set, had installation costs of $2.1 million, and threatened hardware expenditures over the next decade of $37 million. Making the decision easier were predictions of annual savings of as much as $5 million by 1970 and reduced reservation processing times from forty-five minutes to just seconds.

After a fitful start due to inadequate software, SABRE (as it was now called) came on-line in November 1962. It proved to be an extraordinary success, so much so that other airlines attempted to imitate it. The first and most successful was Eastern Airlines, with its APOLLO system (also IBM based). Similar attempts by United and TWA to install customer reservations systems (CRSs) with other computer vendors initially failed, so by 1970 both acquired Eastern software and set up their own APOLLO-type systems.

In 1975, American made the crucial decisions to adopt the Joint Industry Computerized Reservation System (JICRS) standard and to move SABRE into the heart of its corporate business strategy. In doing so, adding the instantaneous delivery of a product (airline reservations) to its traditional business of air transportation, American Airlines became one of the first firms to move toward the production of virtual services.

Soon American discovered that, having paid for the initial start-up cost, the incremental cost of additional services was marginal. SABRE quickly turned out to be an extraordinary money-making machine. Planned return on investment, with all the incremental revenues, was forecast at 67 percent—instead it turned out to be more than 500 percent. Soon SABRE was handling travelers' other needs, such as rental cars and hotel reservations, thus extending its reach beyond the actual flight itself.

One of the most interesting effects of SABRE was that it moved the "factory" closer to the customer. Older ticketing techniques had worked to the advantage of larger travel agency chains, which could implement their own computerized record keeping and, because of their size, get better treatment from the airlines. But with SABRE, even the tiniest agency, if armed with a network terminal, could quickly order the right ticket and an array of ancillary services.

With all of these advantages, it should not be surprising that CRSs have continued to grow in influence and value. In 1988, American's SABRE had revenues of nearly $500 million. That year, for example, American earned $134 million of its $801 million in profits from SABRE—even though SABRE represented only 6 percent of the airline's revenues. In 1989, SABRE was valued at $1 billion, or 25 percent of the company.

By 1990, according to *Business Week,* 93 percent of America's 35,000 travel agents were plugged into CRSs, 12,443 of them with SABRE. And SABRE has continued to expand its services. One new offering is SABREvision, designed to link CD-ROM players at its member agencies with their SABRE PCs and provide full-color images about vacation packages.

One unexpected result of SABRE and the other CRSs is that greater control is placed in the hands of the user—not just travel agencies but corporate clients and even individual consumers. As travel agencies gain the capacity to schedule and book entire vacation packages, and even to print out the tickets, the ability to perform the basic task of making ticket reservations is moving further down the pyramid. According to Calvin L. Rader, CEO of WorldSpan, currently the third largest CRS, his firm wants to become "a true information company" by allowing corporations to use their own desktop computers to perform ticketing transactions.[29] Similarly, software is now available for individuals so they can use their home computers to reserve airline tickets.

An interesting sidelight to all this, and a glimpse into another impact of virtual products, is the customer's reaction to these systems. While CRSs may have begun offering their services to other airlines for covering overhead, in the process they raised customer expectations for a nonbiased presentation of information. Thus, when SABRE and others were accused of internal bias, of diverting users to their airlines or unfairly measuring stopover times for international flights, the result was an extraordinary hue and cry about a breach of public trust. The U.S. Congress investigated, and a bill was introduced (although it eventually failed) to force all carriers to divest their CRSs. American Airlines and the others discovered to their surprise that in virtualizing they had also created a whole new business environment.

Recently, the *Wall Street Journal* reported on new software for travel agencies that more efficiently navigates through CRSs in search of the best ticket prices and even "sits" on the phone line, tracking the status of fully booked flights and waiting to pounce on the first cancellation. With this announcement, the SABRE system, which had originally been American's scheme to capture customers and give it a competitive advantage in the way it presented data on flights, had now become part of a network that would yield control to the customer.

The story of SABRE is a dramatic example of how virtual services can be created and in the process become essential to the conduct of business. Control of the reservations systems has provided American with numerous competitive advantages. Not only has it been a source of great profits, but it has given the company a gold mine of information that has enabled it to run its business more effectively.

For example, the SABRE system can predict (based on historical reservation patterns) whether a flight is going to be full. Obviously, the objective is to carry as many passengers as possible at as high a fare as the airline can demand. Therefore, airlines want to save some seats for last-minute business travelers who pay three to four times as much as vacationers do for many tickets. The trick then is to use real-time information on reservations to adjust fares at the last possible moment to fill the planes. If it looks like there are going to be lots of empty seats, the airline can increase the number of low-priced seats available on the flight.

Using this technique, airlines currently make more than eighty thousand fare changes each day—a number that will only increase with the new travel agency software programs waiting to pounce on the latest fare change.[30] Not only would all this be impossible without powerful information processing equipment, but the competition is so great and the pace of change so fast that the performance of these systems must be perpetually updated. The airline with the best information has the greatest chance of filling the seats on its flights at the highest prices. Over time a few percentage points in load factor can make the difference between profitability and Chapter 11 in an industry that makes less than 1.6 percent profit on sales.[31] For these firms, information and virtual products mean survival.

## Along a Wide Front

The examples selected for this chapter span a wide range of industries. This alone provides some important clues on the nature of the virtual product. The breadth of the examples suggests the eventual pervasiveness of this new business model. If it can affect everything from the manufacture of guns to computer chips, it is easy to envision it affecting a very broad range of discrete manufacturing processes.

There are also trends to be gleaned from the examples in this chapter. For example, the engineering processes described for the semiconductor industry are also being used by other industries, such as in the design of electronic equipment. Mechanical design automation packages from software firms such as Autodesk and Intergraph now enable engineers and architects to design homes, buildings, and airplanes in a fashion similar to that used with electronic devices. And finite element analysis tools permit the performance of mechanical structures to be simulated on a computer just as one does with electronic circuits.

The American Airlines SABRE story suggests how virtual products will change the information-sensitive portions of other service industries. The story of printing also offers clues to the similar changes that will take place in information services. Combine these with the changes in financial services mentioned earlier, instantaneous mail-order fulfillment, and the demand-responsive merchandising systems at companies such as Benetton and Wal-Mart described later in the book and it should be apparent that virtual products are already having considerable impact on traditional service industries.

Another feature of virtual products, an increasing role for the customer as a coproducer, can be seen in the semiconductor, printing, and airline examples. In each case, the customer, by becoming more competent in products and services, can not only better control the result but in the process develop strong and enduring ties to the provider.

The Beretta example offers a telling history of the evolution of the modern corporation and the changes required at each step along the way. Ultimately, it provides a glimpse at the type of manufacturing processes the virtual corporation will employ. The Beretta example also suggests how total quality control and computer-integrated manufacturing will have broad implications in the future of the corporation, as will the creation of a highly trained, involved work force. The Remington story complements this by showing how modern methods offer the possibility of not only advancing the power of manufacturing but enriching it by regaining some of what has been lost in two centuries of mass production.

In the industrialized world, the new business revolution will be all-encompassing. Companies will be moving forward along a wide front, by changing not only themselves but, through interaction involving common needs, other businesses as well.

Making the advance smoother will be the engine that will propel these companies forward: information technology. Information processing makes the virtual corporation possible.

# 3

..........................................................

# Powers of Information

The phrase "power of information" is used so frequently that it has become at best a cliché and at worst an advertising slogan. In the process, the phrase has been so inflated into science fiction—the omnipotent and dangerous supercomputer peering into our lives—that information seems some priceless and enslaving commodity.

The truth is, extraordinary advances in information processing will serve as the dynamo of the virtual corporation. Furthermore, in the years to come, incremental differences in companies' abilities to acquire, distribute, store, analyze, and invoke actions based on information will determine the winners and losers in the battle for customers.

For those unable to previously identify with the implications of the power of information, the events of the Persian Gulf War should have been a revelation. For it was a war in which information gave the Allies the power to win over the bulk of its enemy. After all, solely in terms of available manpower within the zone of combat, the Iraqis at the beginning of the campaign held a 20 percent edge. Most military texts of the last two hundred years have argued that to be successful an attacking force must be twice to three times the size of an entrenched defender. So, by all traditional measures, the Iraqis should have been able to put up a good fight.

Yet, the war itself, first over Baghdad and then on the ground in Kuwait, was among the most one-sided in history. The Allies lost fewer than 100 of their 450,000 troops in the war zone. As has been much remarked since, American troops were statistically safer on the front than they would have been on many city streets at home.

Why the rout? The answer was obvious to anyone watching television during the brief course of the war. The Allied troops had information on the Iraqis far superior to any information held by their enemy and then managed that information more effectively. In strategic command and control, the Allies used satellite and reconnaissance information to learn just about everything they needed to know about Iraq's static defenses. Tomahawk cruise missiles, using reconnaissance information stored in their memories, were so well programmed that they actually followed roads and streets to their targets. Further, because the Allies appreciated the importance of information, they also knew the value of denying it to the other side. They either destroyed or negated the ability of the Iraqis to deliver commands or identify attackers—a plan made unforgettable by the image of Baghdad antiaircraft fire filling the air long after a Stealth fighter had passed.

In battlefield tactics, the power of information was even more striking. While some Iraqi commanders were reduced to using runners to carry messages, Allied soldiers used hand-held digital devices to pick up satellite transmissions and locate themselves accurately within a few yards in the trackless desert. Minicomputers quickly calculated complex artillery trajectories, put shells dead on target, and wiped out the enemy before it could mount a counterstrike. M-1 Abrams tanks used their on-board computers to fire accurately on the fly and wreak devastating damage on Iraqi armor with only a handful of losses.

With the Persian Gulf War, the value of information reached a zenith—bytes became more important than bullets. The case was made, with brutal clarity, that information could become the foundation for devastating destruction.

If there were one piece of Allied hardware that captured and held the world's attention during the short course of the war, it would have to have been the Patriot antiballistic missile. For several days at the height of the air war, Israeli and Saudi Arabian cities were held hostage to Iraqi Scud missiles. Like the German V-2 before it, the Scud threatened the unstoppable horror of a ballistic missile aimed at a civilian population. The jury-rigged Patriot, though only partially effective, saved the day.

Historically, ground defense against airborne attack has always been exceedingly difficult, from antiaircraft fire over Berlin to SAM missiles launched over Hanoi. The intercept times are so short that any deviance in the path of the incoming object is usually enough to guarantee defensive failure. Drop a ballistic missile in from the stratosphere and the population is essentially left undefended and paralyzed in fear.

The Patriot changed that. In some instances, Allied satellites detected launch backfires as the Scuds left Iraq, then transmitted the data to ground stations in the United States and elsewhere, where it was processed and then relayed to Patriot batteries in Israel and Saudi Arabia. There, the surprise gone, the skies could be scanned by radar for the incoming missile, the lock-on information conveyed to the Patriot's own guidance system, and the missile launched. In flight, intercepting at a closing speed of several thousand miles per hour, the Patriot would make the final real-time trajectory adjustments to adapt to the flight path of the often-tumbling Scud. The result was seen in the skies over Tel Aviv and Riyadh.

What enabled the Patriot to accomplish this remarkable task was that it continuously changed its flight behavior to adapt to evolving circumstances. It did this by using sophisticated (though hardly state-of-the-art) computational equipment. But more important was the information this equipment processed, information that captured not only the nature of the attacking ordnance, nor just its overall trajectory, but the second-by-second behavior of the Scud as it descended through the turbulent atmosphere.

In many ways, the ability to predict—whether it was the path of ballistic missiles or the movements of ground troops—and then quickly react was the decisive factor for the Allies in the Gulf War. In the process, time seemed to accelerate, such that the real ground war, expected by many experts to drag out for months, in the end essentially lasted only forty-eight hours. Because the Allies knew so much about the deployment and behavior of the enemy, they could encircle and destroy the Iraqi troops with certain knowledge that there would be no significant resistance.

# Getting the Information Right

One might argue that the power of information in battle is not the same as its strength in other scenarios, for war has its unique intensities and skills. Can information really be of comparable importance in business? The answer is a qualified yes.

Certainly, better information won't allow American Airlines to annihilate USAir overnight. Sears won't vanish tomorrow because Wal-Mart and Kmart use information more effectively. Motorola won't destroy its Japanese semiconductor competitors because it has a better information-gathering network. However, over time better information can translate into a decisive competitive advantage. It can be said with certainty that if two competitors are equal in other respects, the one making the best use of information will undoubtedly win—sometimes with stunning speed.

Consider how Toyota, the world's most successful automobile company, sells cars in Japan. Toyota has shown the competitive power that can be gained by adding electronics to an efficient, low-tech, information-reporting system. *The Machine That Changed the World* describes how Toyota also used computer intelligence to amplify an established information-reporting program in its distribution system.

Like many Japanese auto companies, Toyota long sold automobiles door-to-door, an unusual process outside Japan. But in doing so, Toyota has probably learned more about its buyers than has any U.S. or European car company. Such a sales technique has led to a tight relationship between company and customer:

> Toyota was determined never to lose a former buyer and could minimize the chance of this happening by using the data in its consumer data base to predict what Toyota buyers would want next as their incomes, family size, driving patterns, and tastes changed. . . .
>
> [Sales] team members draw up a profile on every household within the geographic area around the dealership, then periodically visit each one, after first calling to make an appointment. During their visits the sales representative updates the household profile: How many cars of what age does each family have? What is the make and specifications? How much parking space is available? How many children in the household and what use does the family make of its cars? When does the family

think it will need to replace its cars? The last response is particularly important to the product-planning process; team members systematically feed this information back to the development teams.[1]

In recent years, as door-to-door selling has grown too expensive, Toyota has begun a gradual shift toward dealerships in Japan. But even these dealerships are nothing like their U.S. counterparts. Because Japanese car makers are capable of responding very quickly to changes in demand—automobiles can be custom ordered and delivered in a matter of days—dealerships are often little more than storefronts, with a few sample cars.

Here is a typical experience for a Japanese automobile customer:

> The first thing a consumer encounters on entering a Corolla dealership today is an elaborate computer display. Each Corolla owner in the Corolla family has a membership card that can be inserted in the display, just as one would insert a card in a bank machine. The display then shows all the system's information on that buyer's household and asks if anything has changed. If it has, the machine invites the owner to enter new information. The system then makes a suggestion about the models most appropriate to the household's needs, including current prices. A sample of each model is usually on display in the showroom immediately adjacent to the computer display.[2]

That's not all. Using the same terminal, potential customers can also access information on insurance, financing, even parking permits. They can also learn about used cars, if they so wish.

Record keeping on processes inside the firm has been equally extensive. There Toyota uses its famous production system, centered around *kanban,* a simple system of cards attached to part bins that track the progress of those parts through the production process, serving as both a tracking tool and an order form for new parts.[3] In the past, with *kanban,* Toyota could use information management to obviate the need for warehouses and high inventories, escape the roller coaster of over- and underproduction by its suppliers, and implement the first just-in-time production system—and do it all without computers.

When computers did come to the Toyota factories they were smoothly integrated into the process, because Toyota already understood the flow of information in its production areas. Just as important,

Toyota workers were committed to making both the *kanban* and automated information processing systems work. Labor and management worked as a team. This second lesson was lost on many companies, most famously General Motors, which rushed to install computers and robotics before it fully understood the information–worker culture in its plants and the relationship between its factories and dealers.

Within the GM factories, the environment was poisonous to the notion of teamwork and shared information. Says Maryann Keller in *Rude Awakening,* "The workers wanted to do their jobs well, wanted to be competitive, but all too often they were fighting against unbeatable odds to get their jobs done. Every problem became a confrontation; since there was a basic mistrust between labor and management, it was hard to establish a cooperative environment where problems could be solved."[4] Fittingly, it took a joint venture with Toyota (the NUMMI plant in Fremont, California) to show GM how far it had fallen behind in the business it once dominated.

General Motors suffered a similar breakdown in information transfer in its relationship with dealerships. Like other U.S. car makers, GM expected its dealers to accept unsalable cars they didn't want, often by tying them in with delivery of more desirable models. In *The Machine That Changed the World,* the Big Three U.S. automakers are described in the following way:

> To make matters worse, coordination between the sales division and the product planners in the big mass-production companies is poor. While the product planners conduct endless focus groups and clinics at the beginning of the product-development process to gauge consumer reaction to their proposed new models, they haven't found a way to incorporate continuous feedback from the sales division and the dealers. In fact, the dealers have almost no link with the sales and marketing divisions, which are responsible for moving the metal. The dealer's skills lie in persuasion and negotiation, not in feeding back information to the product planners.
>
> It's sobering to remember that no one employed by a car company has to buy a car from a dealer (they buy in-house through the company instead, or even receive a free car as part of their compensation package). Thus, they have no direct link to either the buying experience or the customer. Moreover, a dealer has little incentive to share any information on customers with the manufacturer. The dealer's attitude is, what happens in my showroom is my business.[5]

These weaknesses in information management may not be fatal for General Motors, but the resulting lack of understanding of the needs of the market has contributed greatly to the firm's loss of market share. The danger of this lack of information becomes even more obvious when one considers that the design cycle on Japanese cars is, on the average, more than a year faster than the five-year cycle of U.S. car makers.[6] This puts the Japanese manufacturers in a far better position to track changing market tastes and then quickly design new models to meet them.

The Toyota story shows that computer technology can be used to reinforce already-efficient manufacturing and direct-sales information systems. This access to information about the market in turn acts as an early warning system for the company about any sudden shifts in customer attitudes. As much as any company in the world, Toyota understands the power of information.

One U.S. company that does appreciate the implications of the Toyota story is Hewlett-Packard. When HP decided to bring computer power to the organization and control of its manufacturing, it chose this as its charter: "Getting the right information to the right people at the right time to achieve our business objectives."

What makes this statement remarkable is that Hewlett-Packard is one of the world's largest manufacturers of computers used in design, process control, and inventory management. Yet, there is no mention of computers in the statement, only of the efficient use of information. The implicit message is that any effective means of information transfer is acceptable. If handwritten notes can do the job, use them; if it means the fully networked data processing hierarchy known as computer-integrated manufacturing, use that—as long as it gets the right information to the right people at the right time to achieve Hewlett-Packard's business objectives.

## Bolts of Information

Another example of the way information can transform a business, this time in distribution, is the quick response (QR) movement in the U.S. textiles industry. QR has been driven by fabric giants Milliken and Du Pont and by such retailers as J. C. Penney, Kmart, and Wal-Mart.[7] For instance, Milliken, challenged by offshore competition, set out to obtain the maximum advantage from its proximity to U.S. customers by bringing technology to bear to slash the time between the appearance of a new fashion and its arrival in volume at retailer sites.

A model for these firms was the Italian company Benetton, which had tripled sales to $1.7 billion in just seven years by delivering knitted goods in the hottest new colors seemingly overnight. One way Benetton did this was to invert the traditional method for manufacturing sweaters. Instead of dying the yarn and knitting the sweaters, it knitted sweaters with a neutral yarn and then dyed them to meet market demand. That way, should the market go from blue to green overnight, not only could Benetton respond quickly but it didn't have to be stuck with obsolete inventory. Benetton further accelerated the process by putting into place fast and sophisticated retailer reporting systems.[8]

A study of the U.S. garment industry determined that it took an average of sixty-six weeks to complete the process of converting raw materials to textiles to apparel and then moving it through the distribution channel to the retail consumer. In the fast-moving world of fashion this was slow enough to cost the industry an estimated $25 billion per year in lost potential business. Most of this damage occurred at the retail end, where an estimated $16 billion was lost each year from markdowns, lost potential sales on out-of-stock items, and excess inventory.

The problem resulted in part from a breakdown in the means for passing newly obtained information backwards through the supply chain. Information about a decline in sales of a particular style would slowly worm its way back from the cash register to the buyer, then to the garment manufacturer, and from there to the textile firm and then on to the fiber producer. As orders were canceled at each point along the supply chain, valueless inventory would pile up.

The obvious solution—that of simply hustling the goods through the system faster—turned out to be inadequate for the situation. There would still be too much waste. Instead, a two-way system was needed: in one direction speeding the apparel down the distribution channels and in the other direction feeding back results to each upstream participant.

To see if the theory of a two-way system was right, a research firm set up three pilot runs. The results were staggering. One quick response chain involving Belk and Haggar (jackets and slacks) saw sales jump 25 percent, gross profits 25 percent, and inventory turns 67 percent. A pilot QR chain in tailored clothing involving J. C. Penney, Oxford, and Burlington not only saw sales up 59 percent and inventory turns climb 90 percent, but cut forecasting errors by more than half. Similar results were seen by Wal-Mart, Seminole, and Milliken in tests with slacks.[9]

The QR system, with its overlapping feedback loops of information, had proven itself. Just as important, the participants in the program had learned to trust one another with what was once proprietary information, such as sales, orders, and inventories. Reports Joseph D. Blackburn of Vanderbilt University:

> The first step in a quick response program is to find willing partners. As in other industries, finding partners is challenging because, as an industry spokesman states, "In your traditional relationships, what's out there today is lack of trust. Nobody trusts anybody. Breaking down those barriers is not easy, but that is the key to quick response. There's got to be a lot more communication. In quick response programs or any other program where change must be made, it is harder than the old way of doing business. A lot harder.[10]

Blackburn adds, "Adversarial relationships block communication and inhibit the unfettered flow of information needed for effective supply chains." As with Toyota, only when the informational and cultural systems were in place could electronics be inserted into the program to amplify its strengths.

The key technologies of QR proved to be computer networks, bar code systems, and scanners. Wal-Mart alone invested half a billion dollars in QR point-of-sale equipment.[11] Now a retailer could scan the bar code on a given item being sold and instantly pass the information on that sale all the way back through the chain to the textile maker. Not only was the style of the product recorded, but also information on color and size, and that information served as the underpinning of a fast just-in-time inventory system along the distribution chain.

By the end of the 1980s the QR system had become so efficient—and the level of trust among its participants had become so high—that different players began to use additional technology to tighten the system even further, and in the process approached delivering the virtual product. One such example is described by Blackburn:

In this case, Milliken supplied the finished product, an oriental rug-design area rug, to a major retailer. Milliken made the following offer to the retailer: "You send us the daily orders that you get from the consumer, we will manufacture the rugs and ship UPS directly to the consumer's home." This meant that the retailer could eliminate not only his entire distribution center for that product, but all of his inventory except for display items in the showroom. The retailer responded, "That will work only on regular sales, but commercial sales are such a spike [in demand] in our sales pattern that we've got to carry inventory to cover the spikes." Milliken responded by asking, "What percent are the spikes?" It turned out that two-thirds of their sales were promotional. If they guessed wrong on those two-thirds of their sales, they had excess inventory. Milliken then said, "It's all or nothing. It's either quick response or it's not quick response; we will satisfy your orders. Now you have to tell us when these big spikes are coming. We have got to share that kind of information." The retailer agreed and now Milliken is taking orders on a daily basis and shipping directly to the consumer. The system responds much faster than through the traditional [distribution center] channel and with a fraction of the inventory. Moreover, the retailer's costs were slashed by 13 percent.[12]

## Coordination and Control

The pivotal role of information management in the corporation is something that has occurred within our lifetimes. In *The Visible Hand,* Chandler argues that the modern corporation was built on intensiveness of capital, management, and energy (and with it, material).[13] High-speed information processing will soon have to be added to that list.

The success of a virtual corporation will depend on its ability to gather and integrate a massive flow of information throughout its organizational components and intelligently act upon that information. The very vigor of this process will result in a less capital-, management-, and energy-intensiveness, one of the fundamental advantages of the virtual corporation over contemporary business enterprises. The virtual corporation will significantly alter the traditional internal balance of individual businesses and of the economy as a whole.

It is Chandler who noted that the first management hierarchies appeared in the railroad industry in the 1850s in response to a need to provide and analyze the information that was necessary to run these complex enterprises:

> No other business enterprise, or for that matter few other non-business institutions, had ever required the coordination and control of so many different types of units carrying out so great a variety of tasks that demanded such close scheduling. None handled so many different types of goods or required the recording of so many different financial transactions.[14]

As corporations in other industries began to experience the same complexity of size and scope in their businesses, they too moved to adopt multiple-level hierarchies of management. General Motors in its heyday boasted of nearly twenty levels of management, as did IBM and General Electric.

As these and other companies have struggled to boost efficiency and become ever more market responsive, they have stripped out layers of management and broadened the remaining span of managers' control. For example, in June 1991 GM restructured its giant Chevrolet-Pontiac-Canada Group, by itself one of the twenty largest businesses in North America. It turned three engine groups into one and cut thirteen freestanding parts-making operations to just eight, in the process "pruning away several vice presidencies" and expanding the responsibilities of the rest.[15]

Today the trend in management is to delegate ever more decision making and control to the employees doing the actual work. Computers now gather and supply the information that was once provided by management hierarchies. In the process, training has had to be substituted for supervision. The modern employee is expected to use the information gathered by computer networks and know what to do instead of having to be told. One proven way of doing this has been through the use of quality circles, groups of employees that regularly meet to discuss how to improve product quality and workplace productivity. Such programs teach employees how to analyze and solve quality problems with minimal management supervision.

In some cases, the lines between worker and middle manager essentially disappear. Rosabeth Moss Kanter, editor of the *Harvard Business Review,* has written that as work units become more participative and team oriented, and as professionals and knowledge workers become more prominent, the distinction between manager and non-manager begins to erode. . . . Position, title, and authority are no longer adequate tools, not in a world that encourages subordinates to think for themselves and where managers have to work synergistically with other departments and even other companies.[16]

T. J. Rodgers, CEO of Cypress Semiconductor, has taken a technological approach to the challenge of removing layers of bureaucracy. Starting Cypress with the goal of maintaining a flexible, adaptive organization no matter how large the company grew, Rodgers from the beginning refused to allow a middle management to form. Instead, he installed a computer system to track the daily objectives of every company employee. Such a system, wrote Brian Dumaine in *Fortune,*

allows him to stay abreast of every employee and team in his fast-moving organization. Each employee maintains a list of ten to 15 goals like "Meet with marketing for new product launch," or "Make sure to check with Customer X." Noted next to each goal is when it was agreed upon, when it's due to be finished and whether it's finished yet or not.

This way, it doesn't take layers of expensive bureaucracy to check who's doing what, whether someone has got a light enough workload to be put on a new team, and who's having trouble. Rodgers says he can review the goals of all 1,500 employees in about four hours, which he does each week. He looks only for those falling behind, and then calls not to scold but to ask if there's anything he can do to help them get the job done. On the surface the system may seem bureaucratic, but it takes only about a half-hour a week for employees to review and update their lists.[17]

At one manufacturing company after another, middle management—the most populous of industrial professions a quarter-century ago—is fast becoming an endangered species. Some executives have even grown vehement in their desire to root out what they see as an impediment to success. *Industry Week* quotes Richard C. Miller, founder and vice president of Aries Technology, as assigning middle managers much of the blame for American industry not adapting quickly enough to new technology. Says Miller: "I'm not real impressed with middle managers. Many are [just] hiding out until they retire.

There are times when I would like to take middle managers by the necks, bang them against the wall and ask: 'Don't you realize the company is going to be out of business in five to ten years if you take that attitude?'"[18]

Similar opinions can be heard these days in services as well, such as banking, accounting, and government. Not surprisingly, given this attitude, business is becoming less and less management intensive and increasingly dependent on competence at the worker level. The role of management in the business of the future is bound to decline as computers gather and provide the information that was at one time the product of middle management and as employees become better trained and empowered to make decisions. Information and the power it provides will flow to the worker.

As for the energy intensity of business that Chandler identified as one of the cornerstones of the modern corporation, evidence of its decline is no farther away than your Sony Walkman. The value of the products we purchase is determined increasingly by their sophistication and information content and less by their material and energy content. Nowadays, more powerful products don't necessarily weigh more, they just do more. Their value is a function of their information processing features.

For example, the items that boost the price on today's automobiles include sophisticated audio systems; antilock brakes; electronic controls for fuel, ride adjustment, traction, air bags, and transmissions; and automatic controls for temperature adjustment and seat and mirror position. All are highly prized features and important profit makers for car companies. All are also highly information dependent and add little to the material content of the car.

Home electronic equipment has become better in performance and capability and lighter in weight than ever before. The nineteen-inch, remote-controlled, eighty-two-channel stereo color TV of today weighs only twenty pounds, compared with its sixty-pound, black-and-white, thirteen-channel-dial, monaural predecessor of 1960. Today's midsize color TV sells for about $20 per pound. A 386-type laptop computer weighs only about twelve pounds and sells for about $200 per pound. This compares with just $5 or $10 per pound for an automobile like the Ford Escort. The complexity of the information content is the defining factor.

Much of the added information value in modern products comes from semiconductors. The semiconductor industry itself is one of the most impressive examples of the energy-conservative nature of information-intensive industries—which is one reason why it is so important to a resource-poor nation such as Japan. A typical semiconductor manufacturing plant that produces $1 billion of output consumes just a fraction of the power wattage of a steel plant or an aluminum plant of the same level of output.

In an information-intensive world, supplying information and providing the capability to process it can add billions of dollars to the value of products without requiring much, if any, additional energy use. For example, the business of supplying electronic information is a $12 billion industry; the value of software annually sold in the U.S. with computer systems currently is $52 billion;[19] and information publishing in the United States is an industry of more than $14 billion. Yet all three consume relatively little energy other than that required to cool, warm, or light the offices of their employees. In just this way, the virtual corporation will depend upon the power of information, not electricity, to create value.[20]

One can make a good case that the virtual corporation will be significantly less capital intensive than today's business enterprise. "Lean production is lean because it uses less of everything than does mass production" (see Chapter 6).[21]

When companies successfully implement such systems they make more efficient use of capital equipment, reduce the need for very much working capital, and increase the output per square foot of their factories. The quick response experiment of the textile industry demonstrates how more efficient production, distribution, and retailing can boost inventory turns and thus reduce the need for working capital.

Taiichi Ohno, inventor of the Toyota production system, has been quoted as saying that his current project is "looking at the time line from the moment the customer gives us an order to the point where we collect the cash. And we are reducing that time line by removing the non-value added wastes."[22] By squeezing wasted time out of the system, sizable amounts of working capital will be saved as goods speed their way more quickly through production and into the customers' hands.

That is the impact on manufacturing. One can add that the generation of information-intensive products is less capital intensive as well. One of the purest examples of this is software, the quintessential case of which is Microsoft. The dominant supplier of software for personal computers—$1.8 billion in sales—Microsoft has reached its present size with one of the fastest growth rates in American business history, while still generating the capital it needs internally. The company's public stock offering was primarily a way just to create liquidity for its owners.

Businesses that add value to their products by using information generally enjoy very high gross margins. Take a data base company, such as Lexus/Nexus, a supplier of information to the legal profession. Once it has incurred the cost of installing the computers and collecting the core data, that information can be sold repeatedly with little additional expense. By the same token, the variable costs associated with handling more reservations on an airline reservation system such as SABRE are practically zero.

One of the best examples of an information-intensive product is the Intel microprocessor. These microprocessors power most of the world's personal computers, and Intel enjoys an estimated 80 percent gross margin on each one it sells. Obviously, the value of these products rests not in the cost of their production but in the information processing power they bring to customers.

## Business as Information

So pervasive and fundamental is the role of information—from the words in this book to the strands of DNA in each of the ten trillion cells of your body—that we often look right past it. Most of this information is clustered into data bases, collections of information that range from addresses in a Rolodex to memories in a human brain to the banks of billions of bytes of data supporting a Cray supercomputer. The creation of the virtual corporation will result from linking relevant data bases into ever more extensive and integrated networks.

The data base we most often use during a typical day is stored in our own memory. In it we carry patterns of behavior, records about our friends and business associates, and rules about how to deal with the vagaries of daily life. We augment this body of information with a growing number of other data bases. For example, there are notes we write to ourselves, to-do lists, address books, calendars, and computer data bases. We also transmit this information to others via written and spoken word, delivery systems such as the postal service and Federal Express, and electronic means such as fax and the telephone.

Not only are we receptacles of information, we are also generators. And the information that we generate often induces and controls the actions of others. Take an average workday. An engineer generates designs that encode information in the form of specifications that are transported to others to define the nature and form of a manufacturing process. An architect does the same thing in the blueprint for a building. A stockbroker transmits information and its analysis to clients about market behavior to induce buy and sell orders. In fact, a significant portion of the population that works in service industries does little else but process and manipulate information. Consider the work of lawyers, accountants, advertising account managers, newspaper reporters, and thousands of other service persons.

Here at the end of the twentieth century, four decades into the computer age, it is increasingly obvious that the very nature of business itself is information. Many of the employees in any corporation are involved in the process of gathering, generating, or transforming information. In a typical modern automated industrial firm only a small fraction of the workers—perhaps less than 5 percent—is actually involved in physical work. But even among that tiny population, information has revised the workplace.

In the age of numerically controlled machines and robotics, manufacturing has become yet another information process where machine instructions stored in computer memory describe how a piece of metal should be rolled, lathed, stamped, welded, bolted, or painted. Fewer and fewer human beings are actually involved in the production process. Instead, most workers in manufacturing companies are now involved in producing services. They deal mostly with information. They sell products to customers and service their needs; they are involved in manufacturing overhead processes to ensure that the right products are produced and that the proper materials are readily available; they administer and control the operation; they develop plans and strategies; and they design the products and services they have determined the customer desires. This is as true in the most traditional of

industries—steel, automobiles, railroads, tool and die, and construction—as it is in the most modern, such as computer workstations and
bioengineering.

In service industries as well, major portions of the work force engage
in processing information. In a typical airline, a sizable percentage of
employees are involved in generating and using the information required
to run the business. Similarly, the vast majority of employees in a typical
financial services firm is involved in information gathering, information
processing, and generating of customer services based on information.
The same is true in newspaper, television, and advertising companies.

## The Productivity Dilemma

What we have discussed so far suggests that the rate of progress in
modern corporations depends on the ability of business to generate
and process information. Not everyone agrees, however. As Nobel lau-
reate Robert Solow has said, "We see computers everywhere but in the
productivity statistics."[23] Harvard economist Gary Loveman stunned
computer industry executives at the 1991 Stewart Alsop computer
conference with a similar conclusion about this "productivity paradox":

> I'm here to tell you that after several years, my results have been poor
> and the results of many of my colleagues who have tried similar things
> also remains poor. Poor in the sense that we simply can't find evidence
> that there has been a substantial productivity increase—and in some
> cases any productivity increase—from the substantial growth in informa-
> tion technology.[24]

With tens of billions of dollars spent—enough to put an average of
$10,000 worth of computer equipment on the desk of each U.S. white-
collar worker—how can Loveman be correct? One answer comes from
Michael Borrus, of BRIE: "It simply isn't good enough to spend money
on new technology and then use it in old ways. I suspect that for every
company using computers right, there is one using it wrong—and the
two negate each other."[25]

This suggests that perhaps the real impact of computers is still
ahead of us, to be found in the virtual corporation. Historical support

for this comes from the research of Paul David of the Center for Economic Policy Research at Stanford. David studied an analogous technological breakthrough of the last century—that of the arrival of electricity—and found a similar lag between technology and a jump in productivity: "In 1900 contemporaries might well have said that the electric dynamos were to be seen 'everywhere but in the economic statistics.'"[26]

What David concluded was that the social, organizational, and industrial changes required to pass from one technological regime to the next are so complex and profound as to require a half-century to complete. Thus, just as the early steam engines in England were used to pour water into old water wheels rather than to drive the factory shafts themselves,

> the same phenomenon has been remarked upon recently in the case of the computer's application in numerous data processing and recording functions, where old paper-based procedures are being retained alongside the new, microelectronics-based methods—sometimes to the detriment of each system's performance.[27]

The computer era is now approaching its own half-century mark. We would suggest that just as the dynamo finally moved from being a mere replacement source of power to a central role in daily life, defining its era, so too, with the advent of the virtual corporation, will information processing soon take its proper place as the heart of the new industrial paradigm.

## Levels of Understanding

In a discussion about the power of information, it is important to ask about what kind of information we are talking of. Accuracy aside, all information is not equal. In fact, the information that is of use to a corporation falls into four distinct categories: content, form, behavior, and action. Until recently, only the most elementary category has been available to industry in any systematic and manageable way. Obtaining or generating the other three has become economically feasible only in recent years. Learning how to acquire and work with these other information forms not only will be important but will be the basis of the virtual corporation.

## CONTENT INFORMATION

Content information formed the basis for the modern corporation, as described by Chandler.[28] It is information about quantity, location, and types of items. For example, a simple inventory system includes information about the number of parts of a particular type and where they are located. Other related files may describe the specifications of these parts. Data bases on personnel can contain addresses, employment and salary histories, and even personal data on employees. And customer files will contain data on such items as credit history, past activity, and current orders.

Content information is historical in nature. It records what an employee has done; when an individual was born; where inventory has been stored; what a customer has ordered. Content information is what typically used to reside in a file cabinet and then was transferred to punched cards and now fills the memory of a personal computer, for instance, in a personnel department or on a loading dock.

Until the 1980s, the computer industry was built on the ability to process content information for business. IBM's dominance of the commercial computer business was a direct result of its doing a better job of processing and integrating content information for business than any of its competitors.

## FORM INFORMATION

Form information describes the shape and composition of an object. In comparison with content information, this type of information can be quite voluminous. For example, the content information about a Ford Taurus might describe its color, the type of options installed in the car, the price, and where the vehicle is located in the distribution system. The form information on the same car would describe the shape of every component in the system. It would contain data on the precise shape of a piston, fender, and engine block. Millions of bytes of data are required to store the form data for the same automobile that would be described by a few hundred bytes of content data.

Another illustration of the differences between content and form information can be found in the data associated with a typical American home. The content information on such a house—lot size, square footage, tax burden, mortgage payable—constitutes a few hundred bytes on a tax roll or in the computer at a title office. But to describe the form of that house on a computer system using, say, Autodesk software, would require millions of bytes of storage and millions of computer operations.

Billions of computer operations are often needed to generate the huge data bases that describe form. Twenty years ago this data might have taken weeks, even months, to generate. As computers have become faster it has become possible to generate comparable form data in a matter of hours, even minutes. That compression of time, and its commensurate reduction in processing cost, has made the use of form information feasible for a growing number of businesses.

Content information is a record of the past, and form information describes shape—neither offer much of a glimpse into the future. Yet successful forecasting is a hallmark of business competitiveness, requiring the ability to simulate, often in real time, how a system is going to perform. Such forecasting enables us to understand each system's potential behavior.

### BEHAVIOR INFORMATION

Behavior information often begins with form information and usually requires massive amounts of computer power. To predict the behavior of a physical object a computer must be able to simulate its motion in three-dimensional space through numerous discrete steps in time.

For instance, to simulate the performance of a microprocessor circuit containing one million devices and operating at 100 megahertz (100 million cycles per second) for a single second means the computer must determine the states of one million devices for every one hundred millionths of a second of simulated time. The computers and information processing techniques required to perform these kinds of computations are only now coming within reach. Powerful workstations supplied by companies such as Sun and Silicon Graphics make it possible for engineers to do this type of analysis at their desks.

Computer simulations have also proved to be an extremely effective tool for molecular design. The simulation of a complex organic molecule on a Molecular Simulations system may require anywhere from eighteen billion to two trillion computer operations.[29] In an example closer to the lives of most readers, Boeing uses sophisticated simulation information to study the behavior of aircraft wing designs under stress.

The usefulness of behavior information as an alternative to expensive, destructive, or even dangerous real-life testing has even been recognized in more mainstream industries. "We're at a turning point because of the drastically reduced prices of these [new computers]," George Dodd, head of computer science at the research labs of General Motors, told the *Wall Street Journal.* According to Dodd, engineers will soon be doing tasks that they once only dreamed of, such as crash-testing cars on desktop computers, instead of smashing into concrete walls, and, possibly, designing car parts entirely by computer.

With accurate simulations it is often possible to build a flawless or fail-safe device the first time. Of course this is only practical if the cost of simulating the result and the time to perform the simulations are much less than the cost of building the prototype. The extensive use of behavior information in the last few years has become possible as computer processing power has strengthened and memory storage prices have fallen.

The paradigm of this for our time is Boeing's design of its new 777 airliner, the largest project ever undertaken by computer alone. Costing an estimated $5 billion, it requires seven thousand specialists in two hundred "design-build teams" linked together by seven mainframe computers and twenty-eight hundred workstations. The plane will even be electronically preassembled long before it reaches the manufacturing floor. Despite this mammoth production, Boeing believes computer design will cut the total project cost by 20 percent.[30]

With behavior information, many design disasters of the past might have been averted, reduced to mere laboratory curiosities. Using such information, a potential and unforeseen future tragedy can be replaced with a successful and predictable conclusion.[31]

Further opportunities to use behavior information are waiting in the wings—the so-called grand challenges. These include predicting weather, analyzing atomic structure to create new materials, understanding how drugs affect the body, locating and extracting oil, and understanding the interplay of substances in combustion systems. According to Kathleen Bernard of Cray Research, "Problems such as these can require a trillion [operations] per second and billions of words of memory"—power only a thousand times greater than is available today.[32]

## ACTION INFORMATION

The final triumph of the information revolution will be the use of a fourth type of information—information that instantly converts to sophisticated action. As computing power grows inexorably cheaper, compact, and more powerful, it is possible to build machines that not only gather and process information but act upon the results.

Of course simple feedback machines capable of basic action have been around for years. The standard home thermostat is a good example: it senses temperature and then turns the heat on or off to maintain the house at a constant temperature. Today's action information machines are vastly more sophisticated. As industrial robots, they can accept information and use it to shape mechanical parts, inspect and pick and place parts, or, in a scenario right out of science fiction, build the next generation of industrial robots.[33] In other incarnations, they can evaluate requests for money at an ATM, laser cut a metal shape to order, understand and execute human voice instructions, or build dishwashers at General Electric's celebrated $300 million Louisville plant.[34]

These are but early applications of action-based information. Just as form information and behavior information need both cheap storage and inexpensive processing power, managing action information requires all that came before it, plus inexpensive interfaces to the natural world. This means analog-to-digital converter chips, man–machine interfaces, inexpensive sensors, and sophisticated machine vision systems.

Most of these technologies are already in development. For example, Synaptics is experimenting with techniques to provide machines with humanlike vision—a process that will require not only massive amounts of computing power but organization of that power into brainlike neural networks.

These advances suggest that our lives will increasingly be dominated by machines that perform tasks for us. In many cases this will be a less personal world; the first clues can be experienced now whenever we deal with an ATM rather than a human teller, or when our phone calls are answered by a voice mail machine. Not all of the changes will be so obvious, of course: when we step on the antilock brakes in our car, it is a computer that tells the system how rapidly to apply and release the pressure.

Work environments will also change. More and more factories will operate with "lights out," devoid of humanity except for the occasional passing guard or visiting repair person. One likely effect of this will be a kind of corporate future shock. For hundreds of years, businesses operated primarily using only one type of information. Each generation found new and more effective ways of gathering, processing, and using this information. Now, in just a single generation, three new kinds of business information will be available. Learning to use these new forms will require cognitive changes in both management and the work force.

As resistant to this change as they might be, many companies will find no alternative but to use form, behavior, and action information—if only for the decisive edge they might also give a competitor. With form information, that competitor might be able to eschew the cost and time of prototyping; with behavior information it might better predict the future; and with action information it would be able to run automated factories and provide customers with adaptable products. That is simply too much of an advantage in a competitor for even the largest firm to withstand for long.

While any precise vision of the future may be clouded, it is obvious that the virtual corporation will increasingly rely on devices that harness, integrate, and effectively use these new types of information. Information will be the core of the virtual corporation. A company's ability to operate and create products and services will be dependent on its information-gathering, -processing, and -integration skills. Content information will continue to determine the state of the business. Form information will make it possible to define the products to be produced. Behavior information will test those models in use. And action information will pull it all together into the actual manufacture, testing, and distribution of the finished goods and services. A virtual corporation will be defined by its ability to master these new information tools.

**4**

. . . . . . . . . . . . . . . . . . . . . . . . . . . . . . . . . . . . . . . . . . . . . . . . . . . . . . . . . . . .

# The Upward Curve
# of Technology

The sudden and sweeping transformation of business through technology is nothing new. For example, the locomotive, automobile, airplane, transistor, and integrated circuit have each in their time overturned the status quo. Historically, whenever important technological innovations have resulted in improvement equal to at least one order of magnitude, revolutionary changes have occurred in the way people live their lives and conduct their business.

As we have already noted, perhaps the most famous example of such a change is the Industrial Revolution, beginning around 1770 with the construction of the first textile machines and the first general industrial usage of steam power in England. From that time until 1851, when the Great Exhibition was held in London's Crystal Palace, the productivity of the textile worker, iron worker, and steel worker increased by a factor of about three hundred—or just over two orders of magnitude.[1]

This jump, greater than the combined productivity improvements of the previous fifty generations, was made possible by a burst of inventions—everything from steam power to punched-card Jacquard loom programming to the Bessemer process for creating steel. Not only did each of these inventions make a profound change but their interaction rewove the social fabric. As the great British naturalist D'Arcy Thompson wrote in 1895, "Strike a new note, import a foreign element to work and a new orbit, and the one accident gives birth to a myriad. Change, in short, breeds change."[2] Much of the world in which we live is the result of these early Victorian industrial and social transformations.

Order-of-magnitude advances and their dramatic effects can also be clearly seen in the history of transportation technology. Early man could carry only small loads; thus his principal activities were confined to hunting for or gathering food and constructing shelter. In the ages that followed, the wheel was invented, powerful animals such as the horse, camel, and elephant were tamed, and crude boats were constructed. With those advances, it became practical to carry not just fifty pounds, but five hundred or even five thousand pounds. The resulting surplus made primitive forms of trade possible.

Ancient histories record that (perhaps apocryphally) in 332 B.C. Ptolemy Philopator had constructed a type of catamaran—a four-thousand-rower tesseraconter—capable of transporting twenty-eight hundred soldiers.[3] By 250 B.C. Hieron II of Syracuse had reportedly built a warship weighing forty-two hundred tons.[4] By the time of Christ, merchant ships with five hundred tons deadweight were not uncommon.

Trade, made possible by improved transportation, was responsible for the growth of great cities such as Carthage and Rome. Improved methods of transportation enabled farmers to transport necessary grain to dense population centers that otherwise would have been incapable of supporting themselves. The free flow of goods permitted cities to specialize production, leading to the development of skilled work and guilds. Inevitably, countries once widely separated came into contact with each other and felt the need to protect their spheres of influence.

The ability to move large groups of material and equipment over great distances changed the nature of warfare as well. War was no longer an event that took place almost exclusively between local tribes on a small scale. As early as 490 B.C. it was possible for the Persian fleet to transport one hundred thousand soldiers across the Aegean Sea to Marathon to fight the Greeks.[5]

For the next twenty-three hundred years, however, advances in maritime technology would be incremental, such as with square rigging and the compass. From the trireme to the galleon to the man-of-war, the changes would be only of small degree. The next great change on the sea wouldn't come until a twenty-year span in the mid-nineteenth century, with the arrival of steamships. As one writer put it, "The great three-deckers of 1850 were no different from the carracks of 1450 except for slight modifications in hull shape, a few sail changes, and a certain amount of improvement in the making of firearms. By contrast, a three-decker of 1850 was a completely different vessel from an iron-clad of 1880–1890."[6]

Much the same thing occurred on land. From the chariot and wagon to the Wells Fargo stagecoach and the Conestoga, the rate of progress of land transportation remained glacial for nearly four millennia. Perhaps the only order of magnitude breakthrough that occurred during all those years was the development of the canal, a hybrid of land and water transportation with considerable advantages, although of only finite application. Otherwise, advances in land transportation were limited to improvements in axles, wheels, and suspension systems and to modifications in roads.

Then came the invention of the steam engine and its use in locomotives. According to Chandler, rapid, predictable rail transport made the modern corporation possible.[7] The railroads changed the economic destiny of America. Yet this change could not compare with the social transformation that took place when the speed of personal travel advanced from the six or so miles an hour of a horse-drawn buggy to the sixty miles an hour of an automobile. The car culture that emerged redistributed the population and reorganized the family.

The airplane brought another order-of-magnitude advance in transportation speed. The image of the airplane shrinking the world is a cliché derived from everyday experience. With time no longer a constraint, the average citizen could visit more places in a week than Columbus, backed by a national treasury, could see in a lifetime.

It is important at this point to note that the first order-of-magnitude advance in the speed of travel—from horse to automobile—took more than four millennia to consolidate. The second—from automobile to passenger jet—took less than a century.

In our time, a most dramatic and certainly the most unpleasant jump in technological capability occurred with the development of the atom bomb. Man's capacity to deliver explosive terror to his enemy increased from about 2,000 pounds of TNT per bomb dropped during World War II to ten thousand times that during the war's final week, when on 6 August 1945 the Enola Gay dropped Little Boy on Hiroshima. It exploded with an estimated force of 12,500 tons of TNT.[8] This was enough to end the war with Japan in a matter of days, reorder world power, and grip the world in terror of eminent destruction for a half-century. All this came from an overnight advance of about four orders of magnitude in destructive power.

## Moore's Law

The historical examples cited above strongly suggest that the status quo can be dramatically altered by order-of-magnitude changes. If indeed such advances can occur quickly and unexpectedly, then we should be both vigilant and prepared to react to further extremely rapid rates of change.

In fact, there is one area of technology where order-of-magnitude changes occur regularly every few years; having been doing so for decades, and promise to continue at this pace (or even quicker) well into the next century. The field is information and communication sciences—computers, mass storage, software, and telecommunications. Moreover, these stunning advances can be easily integrated with other emerging technologies to multiply their force. It is this perpetual transformation, achieving technological leaps about once per decade comparable to the four-thousand-year path from horse cart to bullet train, that makes the virtual corporation inevitable and immediate.

As discussed in the previous chapter, the virtual corporation needs vast amounts of low-cost information storage and processing at its disposal to deal with form, behavior, and action information. In addition, it must transmit this information quickly and inexpensively around the world.

The best way to appreciate how all this happened so quickly is to begin with Moore's Law and its implications for business. During the early 1970s, in preparation for a conference speech on the future of memory chips, Silicon Valley pioneer Dr. Gordon Moore plotted on logarithmic paper the capacity of each past generation of computer memory chips.[9] He then plotted the same features for future chips of this type (random access memory, or RAM, chips) planned by his company, Intel Corporation.

To Moore's amazement, the graphed points made a straight line. It seemed that every two years the complexity of these memory chips doubled. He knew his industry had been advancing quickly, but even to Moore this graph was a surprise.

Extending the line still farther, Moore predicted that it would take only the twenty years to 1991 to go from the 1,000-bit dynamic RAM of 1971 to volume production of 1,000,000-bit memory chips. At the time, the very idea of such a powerful chip was fabulous, but history proved Moore correct.[10] Says Denos C. Gazis of the IBM Research Center, "The net result is that we are able to quadruple the density of memory chips roughly every three years. And progress is accelerating."[11]

As it turned out, Moore's Law applied to other integrated circuits as well. Ultimately, this insight has permanently shaped the vision of the electronics industry.

To fully appreciate the sweep of Moore's Law requires a little history. Shockley, Brattain, and Bardeen invented the single-circuit transistor in 1947. In the forty-five years since, that single transistor has been miniaturized and packed with others of its kind into single integrated circuits containing a total of more than thirty million transistors and capacitors. The resulting 16,000,000-bit dynamic RAM represents an increase in capacity of more than seven orders of magnitude. And that is just the beginning. The billion-bit single-chip semiconductor memory is waiting to appear just beyond the start of the twenty-first century, representing a jump of nine orders of magnitude in just over half a century.

Thanks to the advances suggested by Moore's Law, by the early 1970s semiconductor memory had become cheap enough to replace magnetic core memory, the principal storage device in computers. At the time this occurred, core memory cost about five cents per bit. Twenty years later semiconductor memory was available for one-thousandth of a cent per bit, an improvement of four orders of magnitude.

Improvements in semiconductor price/performance have always had a ripple effect upon computers. As Kenneth Flamm of the Brookings Institute told *Datamation*, "It's always been the cost of computing power that has driven computers and their applications, at the low end as well as the high end. Every time the prices tumble like a rock, there's a huge expansion in demand for computers and new applications." Plotting the price/performance of computer central processing units, internal memory, disk storage, and complete systems between 1957 and 1978, Flamm found that all showed a nearly straight-line improvement of 25 percent per year in real dollars—an improvement of one thousand times in just two decades. And Flamm believes the annual gains have been even greater since.[12]

How far the process has gone in forty years is represented in UNIVAC I, the 1950 descendant of ENIAC. It weighed sixteen thousand pounds, had five thousand vacuum tubes, and performed one thousand calculations per second. UNIVAC I was the first commercial computer, though only six were built, three of which were sold to the U.S. Census Bureau at $250,000 apiece.[13] By comparison, a modern computer at the same price, the MIPS 6000, is about the size of a file cabinet and races along at fifty-five million calculations per second.[14]

The computer's information storage capacity underwent a similar leap. Compare a 1961 General Electric disk drive, with its 25,000 bits per square inch, with a Hitachi drive announced in late 1991 featuring 151 million bits per square inch—a capacity improvement of six thousand times in just thirty years.[15]

Comparable improvements have taken place in power consumption as well. That might not seem important until you look inside your home computer and see a power supply about three inches on a side. If it had to be one million times larger, that power supply would be twenty-five feet on a side. With that, the world's current installed base of tens of millions microcomputers would be on the verge of sucking dry the world's electrical power grid. It is the miserly use of power by integrated circuits that makes it possible to pack computer intelligence just about anywhere—on a desk top, in a Patriot missile, in a video game.

## Riding a Tidal Wave

Moore's Law is the most sweeping elucidation of the pace of invention in the electronics era. But there were other, lesser-known, jumps in performance in other areas. Some of the least known but perhaps the most important have been improvements, by many orders of magnitude, in system reliability.

ENIAC, in its day, was almost perpetually out of commission. Its thousands of tubes cooked the interior of the building that held it at 120°F. "Search-and-replace teams combed the machine for blown tubes while other engineers scurried about rewiring major portions of ENIAC to conform to the dictates of each new trial run."[16] By comparison, modern fail-safe transaction computers built by Tandem and others are designed for trillions of computations between breakdowns. This remarkable improvement in reliability has made it possible to use computers in critical automation processes—such as life support, factory operation, and airplane flight control—without the fear of frequent failure.

Although, strictly speaking, one cannot add all advances together because there would be double counting, it can be said that in forty years computing has experienced a combined improvement in five dimensions—mass storage, reliability, cost, power consumption, and processing speed—of thirty orders of magnitude. Such a level of change is almost beyond human compass. It is equal to the jump from the diameter of a single atom to that of the Milky Way galaxy.[17] As we noted above, it took a change of just two orders of magnitude to spark the Industrial Revolution and one of only four orders of magnitude in explosive power to end a world war and redirect human history. Yet, as remarkable as it seems, it will take every one of these thirty-orders-of-magnitude improvements in the world of computing to deal with the form, behavior, and action information required by the virtual corporation. That we are almost there is one of the miracles of the age.

Inventions and their interactions will be the engine of the virtual corporation. A recent study by the Wharton School of Business set out to look at the changing nature of business competition in the twenty-first century and the technologies that would be critical to that competition.[18] Determining that high-speed capital and technology transfer, access to cheap labor, and collapsing product and life cycles would bring about a radical change, the study concluded that "continuous corporate renewal" will be required to remain competitive. It also determined that wealth will no longer be measured in terms of ownership of fixed physical assets but in terms of ownership of (or access to) knowledge-intensive, high value-added, technology-driven systems. "This 'paradigm shift' in management strategy," the report concluded, "will require concomitant shifts in both management structure and operations."

There is a striking similarity between this prediction and the nature of the virtual corporation. The Wharton report goes on to combine U.S. Commerce and Defense Department studies to produce a list of more than twenty critical technologies needed for the United States to stay competitive in the next century. Nearly all depend in some way on information and communications technology.[19]

Most of the inventions in information and communications have occurred and will occur within several basic fields: semiconductors, computer hardware, data storage, software, and data communication. Advances in other technologies—such as laser, xerography, numerical control, speech recognition, computer vision, and liquid crystal and plasma displays—will play an important role as well.

### SEMICONDUCTORS

Semiconductors are obviously vital to virtual products. Within the industry, microprocessors should continue their seemingly endless march up the price/performance curve. Memory chips will continue to get faster and denser, although the rate of progress in this field has begun to fall behind Moore's Law. Application-specific integrated circuits, such as gate arrays, will steadily grow larger, faster, and easier to design—and in the process serve as the building blocks for future generations of computers, numerical control machines, communication networks, fax machines, copiers, cameras, and so on. Thus they will maintain their role as, in chip pioneer Jerry Sanders's words, "the crude oil of the information age."[20]

### COMPUTER HARDWARE

In computers, several new generations of hardware will be needed to make the virtual corporation a practical reality. RISC (reduced instruction set computing), parallel processing (which will let thousands of computers work together on the same problem), and other architectural innovations will contribute to important improvements in cost, size, and processing power. By the year 2000, these developments should permit computation speeds to reach 1 billion (1,000 million) instructions per second in machines costing about $10,000. That's $10 per mip, compared with the current $150 per mip, yet another order of magnitude change. Armed with this kind of power, designers will at last be able to use behavior information more extensively in their simulations—a key to accurate long-term forecasting.

Another area where this processing power will be important is in the presentation of processed information. The human mind can comprehend numerical information more easily as images or graphs than as columns of figures. For that reason, a great deal of the computational power of modern computers is devoted to the visual presentation of information. If the computations can be done quickly enough, this information can even be presented so as to appear three-dimensional and in motion. Industrial Light and Magic uses precisely these processes to create the visual effects in science fiction movies such as the *Star Wars* trilogy.

## DATA STORAGE

The ability to view form and experience behavior is not only dependent on bringing massive amounts of computer performance to bear on the problem but also requires huge amounts of data storage. Even if dealing solely with files of content information, these files can be massive. For example, each year the IRS deals with more than one hundred million tax returns and one billion special information forms and mails out nearly four million letters requesting additional tax payments.[21]

Early computers contained data files capable of storing information on magnetic surfaces at very low densities. The first IBM drum memory, introduced in 1951, was composed of magnetically coated rotating cylinders containing fifty-eight tracks, with one hundred characters per track. The company's first hard disk drive, the RAMAC 350, introduced in 1956, stored 4.4 megabytes on twenty-four-inch platters in a box the size of a washing machine.[22] Today it is possible to store as many as 3.5 billion bytes on a multiple-platter disk drive the size of a paperback book.[23] Most personal computers now come equipped with forty-megabyte disks tucked inside them.

However, although useful, data files of even this size are not large enough to let users work effectively with form or behavior information. For this, an alternative is already on the scene. Optical data storage, which can store as much as two billion bytes of reference data on a single surface, represents an important advance in information storage.[24] It combines laser and semiconductor technology to store massive amounts of data on plastic disks similar to music-grade CDs. The optical disk (or its cousin, the CD-ROM) can also store still or video images that can be combined with text in multimedia presentations that represent a further improvement in ease of use.

81

The notion of billions of bits of inexpensive information at the fingertips of every employee in an organization has dramatic implications for the virtual corporation. For one thing, high-speed manipulation and control of this amount of information is crucial to custom mass production, which in turn makes the virtual product possible. Only with the ability to store and then process the huge volumes of behavior and action information will it be possible to understand and track the changing needs of each and every customer, or to locate and capture prospective new customers and then design products to meet those diverse needs.

The impact of this new storage technology might be even more pronounced in the service industry. For example, a single compact disc would hold not only all the information currently found in any retail catalog but everything else in that firm's year-round inventory as well. Furthermore, photos of products could easily be moved around, and products could be represented in three-dimensional color, be mixed and matched with other items, or even be modified to match the customer's size. They could then be purchased over a computer network for next-day delivery by air.

Although currently there are only a few hundred volumes of such types of compact discs in use—mostly scientific data bases, computer software, and reference works such as encyclopedias—the maturing of the industry has begun. And the corporate world is beginning to take notice. Sun Microsystems, for example, says that it plans to distribute all its software in the future on CD-ROM. Other companies have announced plans to do the same with product catalogs and data sheets. Logic Automation, which catalogs vast libraries of behavior information to help clients simplify electronic system design, is moving to put onto a single CD-ROM what it used to send to its customers as a sheaf of floppy disks.

### SOFTWARE

The typical computer user deals with layer upon layer of software, each in turn converting the user's language into something closer to the cryptic language the computer understands. Were it not for this buffer between the user and the workings of the system, computers would still be the province of only computer scientists.

Early computers had to be programmed in their own internal machine language. Programmers wrote single instructions on programming pads using alphanumeric codes that were then converted into computer instructions. The process was so long and laborious that it was essentially impossible to write the long programs required for sophisticated applications.

To improve this frustrating situation, engineers began developing programs such as FORTRAN (1957) and COBOL (1960) to translate simple English-like programming commands into computer instructions. The success of these early-generation languages led to a proliferation of potential uses for the computer—and, as every computer owner knows, that led in turn to an insatiable demand for more and more computing power and memory. As the amount of data stored in a typical system grew, new techniques had to be found to deal with it, from early flat filing systems to the modern relational data base.

With languages, operating systems, and data management in place, programmers could now begin work on applications programs, software that would bring the power of the computer to bear on specific needs. What followed was the development of an abundance of word processing, spreadsheet, personal information management, and desktop publishing software—and communications networks to link machines together—that continues to this day.

Computers were linked together in networks so that many users could share data and work on problems together. With this networking came the potential to integrate the data generated by one source with that from another. For example, an engineer working on a design might want to feed the information from a schematic circuit diagram to the person doing the layout of a printed circuit board, who in turn might want to forward the results on to manufacturing personnel so that they could purchase the parts and program the production equipment.

Computer-aided software engineering (CASE) tools have made it possible to produce these new applications faster and with a high level of reliability. As a result, the modern computer user has literally millions of lines of code at hand from which to select thousands of different applications programs. In the words of reporter Esther Dyson, "CASE makes it easier and faster to build software, and makes the programs that result higher in quality, more consistent, and easier to change later on."[25]

On the horizon are new generations of software to take full advantage of the increasingly powerful computers of their time. Among the features of this new software, such as that being developed by the new

IBM–Apple joint venture, will be object-oriented operating systems, data bases, and program languages. These and other advances will make it possible for both professional programmers and consumers to harness the hundreds of mips soon to reside on their desk tops.

### DATA COMMUNICATIONS

The increasing capabilities of semiconductors, hardware, data storage, and software discussed thus far are impressive, yet they will not be sufficient for the virtual corporation. Immediate access to useful information will be a hallmark of the virtual corporation.

Computers will have to reach out to other data sources, and their owners will need to pass ideas swiftly back and forth. To reach beyond the employees of a company to suppliers, retailers, and, most of all, customers will require a much better data communications infrastructure in this country than we now have—not just low-speed voice-grade telephone lines linking personal computers but a network of satellites and broadband fiber-optic cable capable of bringing multimedia into every American home. This will be an expensive task—$100 billion, according to one estimate, twice that according to some others—but the technology is already available.[26]

Says William R. Johnson, Jr., of Digital Equipment: "Today fiber optic technology, with a speed of 100 million bits per second, is beginning to be implemented. Continuing this pattern of ten-fold improvements, the next development [before 2000] promises 1 billion (1 gigabit) per second."[27] Data communications is one area where the new business revolution will move into the public sphere, where business decisions will depend upon political ones, and where the entire nation will have to make important and far-reaching choices (see chapter 11).

## Technology and the Virtual Product

The technologies discussed in this chapter, as well as others still to emerge, will make possible the creation of a panoply of virtual products and services. Some of these offerings—desktop publishing, electronic photography, field-programmable gate arrays, desktop shopping, and airline reservation systems—will be almost pure technological creations, empowering the user to create his or her own custom products.

In other cases the technologies will be almost invisible to users, tucked inside corporations to implement fast-response/mass-customization manufacturing systems, used perhaps to make clothes to order, to manufacture cars in seventy-two hours, or to quickly refill a retailer's shelves.

Thus it can certainly be said that technological invention underpins virtual products. But it alone is not sufficient. Desktop publishing and electronic photography, for example, are not new electronic industries. In fact, they are established industries—printing and image making—that have been revitalized and transformed by electronics, just as the ATM is still banking and Beretta, for all its computers and milling machines, is still a gun manufacturer. Computer integration and flexible manufacturing got Remington to the point where it could explore new business options, but it took inventive marketing for the company to look to its own past for further products. Despite improvements and productivity leaps associated with the new technologies, all of them remain subject to the unique characteristics of their businesses.

In nearly every case, the business revolution caused by virtual products will be led by companies in established industries that have recognized the potential of using some or all of the four kinds of information to reconstruct their business. They will be companies that are observant enough to spot a technological crossover threat emerging in some distant industry and race to incorporate it. They will be the ones that develop multidisciplinary inventions themselves. In every case, technology will be subordinate to, not a substitute for, a complete understanding of the market and the business.

But, because many virtual products of the future will be almost solely the creation of technology, there is a danger that business executives will become overly reliant on it. One early victim of this kind of thinking is General Motors, which has wasted its billions on automation and new facilities in the belief that technology by itself could cause a dramatic increase in business productivity.[28]

The purpose of this chapter has been to show as convincingly as possible that technology has advanced to the point where the virtual corporation is now feasible. In fact, most of the advances we see going on around us in the United States are a direct result of technology.

But if technology is no longer an impediment, neither is it sufficient to create a virtual corporation. The Japanese and Europeans, more than Americans, appreciate that fact. They have begun to create virtual corporations not with technology but rather through organizational innovations and redefined business relationships.

There is a danger in believing that technological supremacy is enough to revitalize and keep our corporations competitive. The journey to this new industrial paradigm requires more than just engineering. It will challenge us to rethink the role of every office, every laboratory, and every factory workstation in the company. This time, technical innovation alone will not save us.

# 5

## The Future by Design

n 1894 the Honorable Evelyn Henry Ellis, a wealthy member of the English Parliament, set out for Paris to buy a car—his destination, Panhard et Levassor (P&L). P&L, as well as the world's leading automotive manufacturer at the time, was primarily a manufacturer of metal cutting saws, a classical craft production operation. The workers in the plant were skilled craftspeople who understood mechanical design and the materials with which they worked. The owners of the company, Messrs. Panhard and Levassor, first talked with a customer, developed specifications for the car, and then assumed responsibility for ordering parts and building the final product. Much of the design and engineering of the component parts took place in individual craft shops scattered throughout Paris. When the parts arrived at the shop, skilled fitters finished the design job with metal files to ensure a proper fit.[1]

Almost a century later, when scientists working on the virtual reality (VR) project at the University of North Carolina became involved in the design of their new offices, Sitterson Hall, the experience was quite different. They decided to simulate on a computer system the experience of living in a building that did not yet exist.

When Sitterson Hall was in the planning stages, the VR researchers who were going to work in the multi-million-dollar building after it was completed converted the architect's plans to a full-scale 3D model—that existed only in cyberspace. When the people who were going to spend their days in the building "walked" through the model, many of them felt that one particular partition in the lobby created a cramped feeling in a busy hallway. The architects didn't agree until the future occupants of the building used the 3D model to give the planners a walkthrough. The partition was moved. The building was built.[2]

A few years later author Howard Rheingold visited the newly finished Sitterson Hall, stepped onto a treadmill, and took a "walk" through the same building. "I put the cybergoggles over my eyes, then entered the virtual building and walked through the virtual lobby. . . . I was able to stroll the corridors of an entire building while physically never leaving one small room."[3]

The scientists and architects, as well as Rheingold, had all been using behavior information to duplicate—even modify—the experience of reality. Similar, though less dramatic, versions of the virtual reality experience are increasingly being used by engineers to visit their new product designs. This technique offers one likely solution to the challenge of ever-shorter product cycle times that will characterize the new business revolution. To experience the physical form, to even operate a product that as yet resides only on the drawing board, holds the prospect of all but eliminating the prototype and test-run phases of product development.

Engineering in the virtual corporation will, in fascinating ways, allow designers and consumers to be involved in the production process in a way that both Ellis and those at Sitterson Hall were involved. What Ellis encountered at P&L in 1894 was an older form of what today is referred to as simultaneous, or concurrent, engineering. By definition, this means that everyone affected by design decisions becomes involved in the design process to make sure that the multiplicity of downstream needs (manufacturability, serviceability, market demand, and so on) are met. In Ellis's case this was pretty simple because the craftsperson who produced the parts also designed them, making it very easy to coordinate manufacturing and engineering.

Ellis did something that not many car buyers have done since his time: he played a central role in the design of his own car. He, not some design shop in Dearborn, Tokyo, or Turin, decided upon the features he wanted and made the compromises and trade-off decisions about which would be included in the finished product. He enjoyed, like wealthy men and women in all ages, the luxury of control. In the Sitterson Hall story, many of the people who would eventually work in the building had a chance, albeit somewhat limited, to redesign the building based upon their electronic experience of its use.

Customer control, which lies at the heart of the virtual corporation, is most fully realized in the products manufactured by such enterprises. This, of course, is already going on today. For example, the customer who buys a programmable ASIC from leading American producers Actel or Xilinx designs the desired product and then programs the virtual product. In another case, Otis Elevator has discovered that architects working with design terminals can more easily specify the elevators to be used in their buildings. In this way Otis links the architect into its design process.[4]

The objective of engineering in the virtual corporation is to compress product development time, to shrink the interval between the identification of the need for a new product and the beginning of its manufacture. Accomplishing this can be a demanding and expensive process, but the motivation is clear: corporate survival. Bill Schroeder, vice chairman of Silicon Valley's Conner Peripherals, one of the fastest growing companies in history (zero to $1 billion in four years) explains it this way: "The first guy cleans up, the second does OK, and the third guy barely breaks even. The fourth guy loses money."[5]

While this may be oversimplification, there are enough companies who have experienced the pain of being late to market that they have decided to do everything within reason to cut development time. They find that products with short design cycles are more successful—they can be designed to respond to competitive threats more quickly, and they more closely match the needs of the market. In addition, development costs are less. That is why Intel, to maintain dominance in microprocessors, has chopped the average development time for its new products by more than 50 percent, from 108 weeks to 48.[6] It is also why the most sophisticated Japanese have cut the design time for a new car to just three years (compared with five years for U.S. automakers) and are working to make even greater reductions.[7]

## Serial Problems

Today the majority of companies worldwide use a feed forward, or serial engineering, design process: Information on customer needs is gathered and a product specification is developed. Engineering then designs the product and passes it on to manufacturing to build. The product is then sold to the customer and a service organization is given the responsibility for repair when the product breaks down. The process is linear—each step must be completed before succeeding steps can continue.

Besides wasting time, another serious problem with serial engineering is that often the upstream participants in the process have little appreciation of the growing wedge of downstream consequences created every time they make a simple decision. The problems caused by this process were evident at Cadillac when it was using a serial engineering process:

> Cadillac used a disastrous serial-design method. The designer of the car's body would leave a hole for the engine, then the power-train designer would try to fit the engine into the cavity, then the manufacturing engineer would try to figure out how to build the design, and finally the service engineer would struggle to invent ways of repairing the car. The results were predictable. On one model, the exhaust manifold blocked access to the air-conditioning compressor, so seasonal maintenance meant removing the exhaust system. On another model, the connection between the spark plugs and the spark plug wires was so tight that mechanics tended to break the wires when they pulled them off to check the spark plugs.[8]

At networking equipment maker Cisco Systems, one of Silicon Valley's fastest growing new companies in the late 1980s, serial engineering created a number of disastrous problems just when the company was doubling sales annually. Poor communications about the specifications for a communication chip resulted in low yields and field failures. Because of inadequate programs for testing its two-interface system, the company was forced to ship four interfaces to customers who wanted just two. Reported *IEEE Spectrum:* "Then there was the manufacturing 'bone pile'—the stack of circuit cards that had failed tests during manufacture. Engineering had created only meager diagnostic tests for these rejects. As a result, manufacturing could not find the cause of many of the failures and could not correct them—and the bone pile got bigger." Cisco eventually solved the problem, and finally escaped it altogether in a subsequent product by moving away from serial design to a more cooperative relationship between design and manufacturing.[9]

## Concurrent Engineering

Obviously, there has to be a better way. Too much is at stake to continue to develop products by poor methods. One study has suggested that 75 percent of a product's cost is decided at the conceptual design stage, that a 50 percent cost overrun on development cuts profitability by 3.5 percent, and that a six-month delay in getting a product to market slashes profitability by 33 percent.[10] As noted earlier, another study has estimated that the potential gains from improving a company's development process are between 40 and 60 percent.[11]

Numbers like these have provided powerful incentives for rethinking engineering and design. Such reevaluation has resulted in the development of a process mentioned earlier—concurrent engineering. Some observers have suggested that perhaps it is less a new methodology than a simple application of common sense. The central notion behind concurrent engineering is that everyone affected by design—engineering, manufacturing, service, marketing, and sales personnel, as well as suppliers and customers—should participate as early as possible in the design cycle. Through this communication, trade-offs can be made and consensus reached, which will reduce product cost, improve manufacturability and serviceability, and ensure that the product's features match the market's desires. By predicting and eliminating problems before they appear and carrying out activities in parallel, months can be trimmed from the design cycle.

Not surprisingly, among the components of concurrent engineering are such concepts as continuous improvement (*kaizen*), just-in-time delivery, total quality, statistical process control, and "design for" manufacture and assembly. The latter is especially interesting because it calls upon designers to constantly keep the needs and limitations of the factory in mind—to produce "robust" designs that manage to achieve better product performance and high quality while being easy to manufacture.[12]

Using a concurrent engineering approach, the Japanese have reduced the time it takes to make production-ready dies from two years to one. In the automotive industry, die makers would often wait for product designers to provide complete specifications before the steel blocks would be ordered and the production of the dies would begin. In the Japanese system, die makers are given a rough idea of the size of the new panels even before the design is complete; close communication between the designers and the die makers allows the latter to begin production while the body design is still in the works. Rough cuts can thus be made early and the final machining follows as soon as the designers are ready. This pragmatic approach saves costly time.[13]

Chrysler learned from the Japanese model when it set out to create the Viper. Using concurrent engineering, a development team of just eighty-five people created the sports car in only three years—two years faster than the typical Chrysler—with a development cost that, at $70 million, was nearly half that of the Mazda Miata.[14]

General Motors also has adopted a form of concurrent engineering for its new Impact electric car, forming a team of young design engineers, manufacturing personnel, marketing specialists, and even production people working alongside one another in an attempt to design the car in just four years.[15]

Ford got its own lesson in concurrent engineering when it developed a minivan jointly with Nissan. As described by a Ford executive:

> It was claimed by Japanese economist Takahiro Fujimoto that Nissan engineers had difficulty in communicating the stage overlapping [concurrent engineering] concept to Ford personnel. . . . If, for example, there were shortcomings at the design stage, Nissan would practice the *toriaezu* approach (or "it's okay for now") and then provisionally proceed to the next stage. Whereas, it was alleged that Ford engineers would lose time by rigidly adhering to 100% qualification at each stage serially, before proceeding.[16]

When teams work in parallel and communicate well they can often begin working before upstream groups have finished the design work. Many of the mistakes that so frequently plague development programs can be avoided by doing things right the first time. Consulting group

Pittiglio, Raabin Todd & McGrath surveyed several hundred midlevel managers and others involved in product development and came up with some surprising results. Among their findings: 47 percent of all product development is repeated and 51 percent of development activity consists of "fire fighting" or unplanned activity; also, project hand-offs are often botched because of poor communication.[17]

Another benefit of using concurrent engineering is that it leads to products that are easier and less costly to manufacture. One remarkable example of this can be found at Volkswagen's Hall 54 assembly plant. As a result of simultaneous engineering, an extra frame part was added so that the front end of Golfs and Jettas under assembly could be temporarily left open. This permitted engines and shock absorbers to be installed by robotic hydraulic arms. Until then, this operation had been the bottleneck of the assembly line, requiring as much as one minute and several workers to complete. Using the new assembly process, the time was cut to twenty-six seconds, unattended, thus enabling the entire production line to run more efficiently.

In another example at the same plant, the assembly management convinced the purchasing department to buy cone-point screws, even though they cost 18 percent more than the standard flat-tip versions. The reason was that the new screws would insert easily into holes even if not precisely aligned—crucial to the application of automatic insertion tools. The end result of these and a number of other decisions was that Hall 54 was able to use robots or special machines to perform 25 percent of its operations, whereas in the past, 5 percent had been the best achieved.[18]

As this suggests, the design of a product has a definite influence on whether it can be flexibly manufactured. A well-conceived design will support various models in production by using large numbers of common parts, identifying subassemblies that express the model differences, concentrating model differences in as few parts as possible, and designing assembly sequences that permit the product to be made in modules.

## Computer-Aided Design

Computers often play a decisive role in making design improvements. The greater the capacity to visualize the finished product, the more likely it is that problems can be spotted early. It is estimated that only

20 percent of product-quality defects come from the production line, whereas 80 percent are "locked in" during the design phase.[19]

Using a computer to attack design problems with an eye on the ultimate manufacturing process can yield remarkable results. For example, NCR's Cambridge, Ohio, facility used three-dimensional computer-aided design software to develop a new point-of-sale terminal. After a review team critiqued the design for shortcomings, the final product had 85 percent fewer parts and required 65 percent fewer suppliers and 75 percent less assembly time.[20]

Digital Equipment Corporation used the same technique in designing a three-button computer mouse, cutting assembly time by 65 percent and materials cost by 42 percent. In a similar program with a new computer storage system, DEC found it had increased the product's reliability by 50 percent over its predecessor and reduced its cost per million bytes by 50 percent.[21]

Thanks to simulation, consistency, and the sharing of data among concurrent work teams, computer-based design can also lead to unprecedentedly high levels of product quality, even as manufacturing performance undergoes manifold improvement. In 1988 Intel instituted its $(PDQ)^2$ (Perfect Design Quality, Pretty Darn Quick) program to accelerate design cycle times. In three years, thanks to computer-based concurrent engineering and improved communication among design teams, the time from design to sample was cut in half, while product complexity still doubled. Most important, the company achieved a 95 percent success rate on the first silicon fabrication of new products. As a result of that success, Intel, in the face of a dozen Japanese competitors, retained 95 percent of the so-called flash memory market, its devices ten times more reliable than any other.[22]

Clever design has been used in many fields to get great product variety simply. In electronics, many new products are merely steps put into programmable memory for a microcomputer to execute—change the memory and you change the features. By the same token, clever design can also keep high-volume products cost-competitive. For example, in the early 1980s Sony automated its Walkman assembly plant by first redesigning the products so they could be assembled from only one side and then building the assembly machines to do it.[23]

The future promises to bring multimedia computing to the design process, enabling concurrent engineering to truly live up to its title.

The first of these products are just now being introduced. For example, CIMLINC, an Illinois software firm, has designed a product that enables engineers and manufacturing personnel to interact on computer screens—using full-motion video and data base texts, creating sketches, and even making balloon notes or arrows on existing, on-screen design drawings. In 1991 one user of this system, a Raychem machining plant in Richmond, British Columbia, was awarded the *ComputerWorld–Smithsonian* manufacturing award for producing zero-defect products nearly four times faster than a job shop and for having reduced set-up times from two hours to just fifteen minutes. Said the general manager of the plant, "There's only one piece of paper in the entire operation: a shipping tag."[24]

## Thinking Ahead

When the design of a product and the manufacturing process used to make that product are well thought out from the start, it is much easier to achieve incremental product development. This in turn lets companies get new products to the market more easily. As one might expect, the Japanese have adopted this philosophy quite broadly in their design efforts. Incremental product development fits quite neatly with the Japanese process of continuous improvement, *kaizen*. Most Japanese companies actually begin their development cycle with a product already in hand before adding somewhat minor improvements. "Over 90% of the product development work in Japan is of the incremental type," explains Tatsuo Ohbora, a principal at McKinsey & Company's Tokyo office. According to author Lewis Young, what sets Japanese product development apart from that of other countries is that this improvement is constant and always strives to tailor the product to individual needs.[25]

A remarkable example of incremental development can be found in the thirteen-year story of a humble, three-horsepower residential heat pump manufactured by Mitsubishi. As Stalk and Hout describe it, when Mitsubishi introduced the first of its models in 1976 the United States was far and away the world's largest maker of residential heat pumps. For three years, Mitsubishi made few changes in its product besides some variations in sheet metal. Then, beginning in 1979, the company began a process of making at least one major improvement in the pump every year through 1988. In 1979 it was the addition of remote control. In 1980, a major breakthrough occurred when it added integrated circuitry for control and display. A year later, the company added microprocessor control and quick-connect freon lines—not to improve performance but to cut costs by circumventing distributors and selling directly through appliance stores.[26]

Each year, the improvements relentlessly continued: a rotary compressor, expanded electronic control, optical sensor control, hand-held remote control for temperature and humidity, circuitry that allowed the product to defrost itself at the appropriate time, electronic air purifiers. By 1989, without a single technological breakthrough, Mitsubishi was building a residential heat pump with twice the performance of its 1976 ancestor and offering a banquet of extra features. Meanwhile, its U.S. competitors were still just beginning to install integrated circuitry. Finally, the leading U.S. heat pump maker gave up and began sourcing from its Japanese competitors.[27]

Perhaps the most fascinating aspect of the revolution in engineering involves the automation of the design process. The key to this automation has been improved computer performance and ingenious software that most notably deals with form and behavior information. The large, lumbering computers of the 1960s were used principally in data processing applications such as accounting, payroll, materials requirements planning, and production scheduling—that is, content information. They could be used to maintain parts lists and records on what was built but were ill equipped to play much of a role in design.

Even without computers, engineering design has dealt with form information for many years, through drawings, clay models, and engineering prototypes. A typical engineering bullpen at an aircraft company during the Second World War consisted of row upon row of draftsmen hunched over boards drawing cross sections of aircraft fuselage and wings, like slices in a loaf of bread. The assumption was that the surfaces being described varied smoothly across the width of these slices. A similar process was used in the design of automobiles. Unfortunately, it is difficult to visualize finished products by looking at engineering drawings. Key features are often missed and the paths of parts moving in three-dimensional space are rarely described adequately.

The way that firms dealt with this problem of translation was by creating physical simulations (usually to scale) of the planned products. For example, automotive designers carved full-size clay models to see how the final product would look. Some kind of prototype was almost always built before production began. With any luck, this would enable designers to discover problems with designs and modify them before it was too late. A typical process would consist of drawing a design, building a prototype and testing it, building a pilot run of a number of units on the manufacturing floor, and then testing those units as well to make sure the design could be fabricated and would work as planned. In the case of airplanes, this meant brave test pilots risking their lives trying out new designs.

An important process running parallel to all this was engineering analysis. Here even the early computers could help. Engineers would describe as much of the design as they could using equations and then run those equations through a range of variables to see if the design held.

While this type of analysis was capable of identifying many problems, in general it was not very accurate and was inadequate for most of the complexities of the real world. Despite these limitations, the market pressure for ever-higher levels of performance constantly forced designers to be more aggressive in their designs. Sometimes, when the limits were pushed too far, the results were disastrous. For example, engineers who designed the Lockheed Electra did not reinforce the wings enough so that when the plane encountered turbulence, harmonic vibrations would cause them to rip right off the plane. This euphemistically named flutter problem caused several horrible crashes before it was identified and fixed.

A similar harmonic vibration problem caused the collapse of the Tacoma Narrows Bridge in Washington. Built in 1940, this 2,800-foot span was so delicate in design that it was called by some critics the most beautiful bridge in the world. Motorists who had to drive it, however, called it Galloping Gertie for the way it rocked in even light breezes; sometimes the car ahead might even disappear momentarily from sight as the bridge bounced in the wind. One morning four months after the bridge opened, a wind came up at forty miles per hour—the bridge began to sway and writhe with increasing severity until it simply tore itself apart and fell into the water below.

The Tacoma Narrows Bridge had been designed using the state-of-the-art engineering analysis tools of its time. Unfortunately, the tools were inadequate for the task of dealing with massive and complex computations needed to predict the performance of the bridge in a storm.

One common way of coping with the lack of precision was to simply overdesign products. A classic example of this is the DC-3, statistically the safest and most durable airplane ever built. The plane was constructed with such a large safety margin (notably in the multispar wing and tail construction) that it had to compromise both performance and payload. On the other hand, the plane could fly seventy thousand hours without being rebuilt—and many of the nearly sixty-year-old planes are still in daily service throughout the world.

Until the advent of high-performance, low-cost computers, design analysis was necessarily crude. Engineers frequently used what were called worst-case analysis techniques, which studied the performance of designs at the most extreme limits. While these techniques often led to reliable designs, they also often unduly restricted the flexibility of the designer. Worse, even with this defensive technique, it was nevertheless impossible to imagine, much less explore, all the potential worst-case scenarios. The only real answer was to build a prototype and then test it to destruction in the hope of finding some undetected or unconsidered flaw.

For those early designers, about the only advantage was that the products were far simpler than they are today. When first introduced, the DC-3 carried only twenty-one passengers, flew just 192 miles per hour, and had two 900-horsepower Wright Cyclone engines. That's a far cry from the modern four-hundred-passenger Boeing 747 or supersonic Concorde.

By the same token, the most sophisticated integrated circuits of the early 1970s contained five thousand transistors—today, that density approaches five million. It was difficult enough to coordinate the efforts of a small team doing the design and analysis of that earlier device. For the modern integrated circuit to be designed the same way would require the coordinated efforts of thousands of engineers, truly a Herculean task.

The electronic design process probably has advanced further in electronics than in any other engineering field. This is partly because electrical circuits are comparatively well behaved and obey simple rules. Also, the histories of computers and semiconductors are so intimately linked that the resulting interaction creates considerable synergy. Electrical engineers who designed chips for computers understood a lot about the potential of those computers to help them do their work.

One of the first uses of computers in design was in the automation of the drafting process. The first work on a computer-driven display of a CRT took place in the early 1950s on the Whirlwind I computer at MIT. By 1962, in a doctoral dissertation, Ivan Sutherland introduced the concept of interactive computer graphics—and within a few years designers at General Motors had turned the idea into a reality. Leading that team was Dr. Patrick Hanratty, who in the early 1970s wrote the first mechanical drafting software, the philosophical core of many CAD systems today.[28]

It was easy to have computers draw lines, rectangles, and circles under the direction of a draftsperson. If a dimension needed to be changed or a line moved, the computer merely erased the form data stored in its memory associated with the old shape and replaced it with new data. Just a few years earlier, the electric eraser was considered an innovation in drafting; after the advent of the computer, almost overnight the drafting table, t-square, and paper became superfluous.

As in most applications of computers, what began as a replacement for traditional methods soon expanded into previously unimagined new uses. For example, since many of the shapes drawn were used over and over again, it was a simple matter to store them in a computer. The draftsperson could then point to a screen at the place where he or she wanted the shape to appear and the computer would draw it. By the same token, the computer could also enlarge and shrink the object on command. *Industry Week* reported that "since this was the first wave of automation to hit design engineering, the results were spectacular. For example, productivity ratios (which compare the time needed to do something the new way versus the old way) were reported at 4:1 to 20:1."[29]

The first systems automated the more mechanical processes in electrical engineering. They helped the draftsperson a great deal but did little to leverage the skill of the engineer. The next step was to develop computer-aided engineering (CAE) tools.

At the time, engineers would prepare schematics or abstract representations of what they wanted the draftsperson to draw for them.

Automating this process was straightforward: enable the engineer to enter the schematics directly into the computer and then have the computer automatically draw the physical representation of the design.

Engineers of course frequently used certain structures over and over again. So almost immediately they began to store on their computers libraries of these structures. Consequently, if an engineer wanted to use, say, an "adder" design in a project, it was a simple matter of accessing a copy of the adder from the computer's library and inserting its schematic into the design. If a microprocessor needed to be integrated into the design, it was borrowed from a schematics library and dropped into its proper place. As computer power grew, designs consisting of thousands or even tens of thousands of transistors could be used over and over again. The designer now merely had to use a mouse to point at an icon representing, say, the circuitry for a modem, click on it, and then place that design into the system, the computer making all the necessary interconnects. A process that once would have taken months now could be accomplished in minutes.

Reusable engineering, as this type of phenomena came be to called, was a boon to designers. But serious questions remained, most notably: How could you be sure the design really worked the way you thought it did? In such complex structures it was easy to make a mistake in the interconnections and severely compromise the resulting product.

The only way to guarantee that this wouldn't happen was to build prototypes and then test them under a wide range of operating conditions. Needless to say, this defeated the purpose. For one thing, building prototypes was a slow and expensive process. In a traditional manufacturing facility, it could take months and thousands of dollars. Then the engineering test set-ups would add even more time and cost—and these tests were still only capable of testing the devices in a limited number of situations.

The obvious resolution was to let the computer do it, let it create realistic models and put them through their paces. A great idea, but only if the computer ran fast enough and was cheap to use. That didn't occur for general business use until the 1980s and the arrival of engineering workstations from firms such as Sun and Apollo.[30]

With workstations, high-speed simulators could be built that reproduced the actual electrical characteristics of devices in different configurations. These simulations were used to build models of even more complex collections of devices, adding to an ever-larger library for designers. At the same time, the continuously improving price/performance ratios of computers, combined with new engineering software from companies such as Autodesk and Versacad, made workstation-like activities possible even in personal computers. As reported in *Industry Week*, "Before PC packages, a mainframe implementation could mean $70,000 per [engineering station], but now the total cost—PC plus software—can be less than $10,000. Even a deluxe PC system would be hard pressed to exceed $20,000."[31]

## The Silicon Compiler

One of the keys to computer-aided design has been the empowerment of engineers to deal with higher and higher levels of abstraction. In practice, this means that the computer is able to present the designer with intuitive, menulike choices while simultaneously taking care of all the tedious details of translation. For example, an engineer might point on the computer screen to an icon representing a microprocessor, a high-level abstraction, and the computer would in turn generate the tens of thousands of rectangles and polygons needed to implement the design. The development of this capability has come from yet another corner of computer technology: the compiler.

Compilers are programs that greatly increase the productivity of software engineers by generating many lines of recondite machine language code for each line of English-like compiler code written by the engineer. Compilers of this type have been around since the 1960s. The great design breakthrough in silicon engineering came at a conference in 1979, when a CalTech graduate student, David Johannsen, working under Prof. Carver Mead, announced the first silicon compiler, software that made it possible with simple language requests from the user to pull together complex integrated circuit design components into larger structures.

Silicon compilers are extraordinarily complex structures. Author George Gilder compares the process to designing a complete town—buildings, homes, streets, water, and power—for several thousand people:

> The crux of compiler design is to partition the overall job into a group of problems of manageable size, in which each part consists of a countable and intelligible number of entities. Making one brick, or one brickmaking machine, or one room design, or one house design, or one town's street design, are all manageable problems. So is designing a transistor, a logic gate, a functional block, a computer architecture. . . .
>
> Silicon compilers perform the crucial function of allowing the designer . . . to avoid the multifarious intricacies of lower-level implementation. . . . Freed from the details, the programmer can address the specific needs of the user.[32]

Silicon compilers and other related tools have made it possible for engineers to take advantage of the tremendous complexity that can be placed on a single silicon chip. It is now possible to design circuits containing millions of transistors, accurately predict their performance, release complete production tooling, and get a part working perfectly the first time—all in a matter of months.

When Moore's Law first forecast the future of the semiconductor industry as a doubling of complexity every two years, there was much handwringing in the business because no one could envisage how to cope with the design complexity required for this rapid rate of advance. But with silicon compilers, computer-aided design and engineering software, ever more powerful computers, and increasingly sophisticated graphics, the design tool industry has managed to close the gap.

Some of the more sophisticated ASIC companies are already involved in this process. For example, here is a glimpse of what's happening at LSI Logic:

> We've now got this extraordinarily rich computing environment, all on $20,000 computers networked together ten or fifteen in a group. . . . And, of course, software methodology and technology has matured greatly and now we are able to create almost astounding virtual reality on these computers—we literally now are capable of modelling and designing and developing complex circuits where the interactive design is performed between the software and the engineer in such a way that can predict the circuit's performance behavior. . . .
>
> [That's how] we've been able to design something like ninety prod-

ucts in less than two years—we're building at the same time we're designing. And virtually all of them work the first time. It's the methodology now that's exciting. Products are a fleeting thing. They come and go in nanoseconds compared to the old paradigm of products living for three years. . . .

We redesigned the MIPS R2000 [RISC processor chip] from the ground up, and it took only ten months from definition to working silicon. That's 700,000 transistors. Ten engineers, ten months, and the exciting thing is that today we're still shipping Revision A, with no known bugs. The first silicon that came out of the fab worked. . . .

The biggest savings in time to market is in verifying that the thing's going to work before you commit to silicon. That's because you take a 12-week hit automatically when you start over. You also get greater reliability this way. We had a major customer here recently. Silicon Graphics. They called us up—this was November—and said, "You know those prototypes of ours you've got coming out? Well, go ahead and just ship them to us in volume, because we can make a million dollars a day if we can ship those boxes before the end of the year."

And we said, "But you haven't signed off on the prototypes."

And the guy said, "Hey, your stuff always works. Just go ahead and ship."

The bottom line is the parts did work and Silicon Graphics came over here in eight stretch limousines and took the whole group out to a fancy dinner as a show of appreciation.[33]

## Higher Levels of Abstraction

Engineering in the virtual corporation will not be the plaything of electrical engineers alone. Designing mechanical entities is not fundamentally different from designing electrical ones. Computer-aided engineering teams already work on the design of airplanes, automobiles, and buildings. Finite element analysis tools enable designers to predict the behavior of mechanical structures in the same way electrical engineers can forecast the performance of electrical ones.

One of the most widely used computer programs for mechanical design is produced by Autodesk. This inexpensive software, which runs on a personal computer, can be used in everything from the design of office space to simulated racquetball games. The company's Cyber-space program, in which the operator wears helmet and glove, "puts the designer in space with the parts he is designing. By moving his head, he changes the view, and his gloved hand gives him an ability to grasp and move parts of the design and grab commands in the heads-up command system."[34]

Few human activities are as clearly defined as architecture. But even in other disciplines, computer-aided design, though more limited, can still be a great aid to human creativity. One such field is that of molecular design, highly dependent on the advances made in computational chemistry and empirical data gathered through X-ray studies of molecular structures and many other techniques. Here, computer-aided designs might seem out of place. Yet one project, at the University of North Carolina, involves using virtual reality to enable chemists to see molecules floating before their eyes, hold them and feel their attraction and repulsion with other molecules, and even bond them together according to the rules of physics to produce new hybrid designs.[35] In addition, similar research is being conducted in performing simulated surgery using three-dimensional X-ray images. Research is even focusing on manipulation of the flow of electricity through integrated circuits.

All of this suggests that the automation of the engineering process has just begun. For example, since 1986 in Japan a hush-hush project backed by two hundred businesses and led by such firms as Nissan, Hitachi, and NEC has been at work on a robust design system that would automatically adjust the dimensions of all the parts in a device to compensate for a single one not conforming to specifications, thus achieving perfect quality.[36]

According to management expert Peter Drucker, some Japanese firms also are taking advantage of the time and cost savings of computer-assisted design to take concurrent engineering to a new, higher level—that of developing three new competing products at the same time. Engineers use their method of *kaizen* to improve an existing product through better performance, greater reliability, and reduced cost. A second program attempts to create a new product out of the old, as the Sony Walkman was created out of the tape recorder. Finally, a third program searches for a genuine, breakthrough innovation. Says Drucker:

Increasingly, the leading Japanese companies organize themselves so that all three tracks are pursued simultaneously and under the direction of the same cross-functional team. The idea is to produce *three* new products to replace each present product, with the same investment of time and money—with one of the three then becoming the new market leader and producing the "innovator's profit."[37]

In the United States, where reusable engineering was first developed, researchers such as E. A. Feigenbaum and R. S. Engelmore of Stanford University now envision the day when each design engineer will have his or her own so-called electronic associate. Such an associate would consist of a powerful set of tools and data bases that would raise engineering to ever-higher levels of abstraction while increasingly simplifying and automating the process. Included in this vision are massive data bases containing not only textbook facts but empirical knowledge related to design, reliability, safety, manufacturing process, and so on, as well as suites of seamlessly integrated engineering tools. Ultimately, Feigenbaum and Engelmore envisage design teams in virtual corporations in which some of the members may be computers instead of people.

[The engineer's associate (EA) would] contain a model of the engineering process from one or more perspectives, and know how to access the knowledge bases and the tools for assisting the various members of the engineering team. In the short term, the EA would mediate among the human designers, help resolve conflicts, and transform perspective. [It] would include automated systems as members of the engineering team.[38]

It almost goes without saying that with the arrival of the engineer's associate and comparable artificial intelligence systems, the very nature of the design process will change. With expertise transportable, it will become possible for customers, acting as coproducers, to assume more of the design process themselves. By speeding the process, engineers will be able to evaluate many more alternatives before selecting a solution. This is precisely what happened at Canon when it installed the OPTEX system for designing zoom lenses. By cutting design time by more than one order of magnitude—from three hours to fifteen minutes—the designers were able to investigate and test many new ideas that had previously been too time-consuming and costly to consider.[39]

105

* * *

The world of the engineer in the virtual corporation will be one powered by high-speed computation and all its wonders. It will be one of organizational innovation in which concurrent engineering and common sense will be responsible for many of the changes. Companies will have the capability to rush products through the design cycle and into the market with breathtaking speed. The powerful tools will also simplify the design process, making it possible for customers to be more deeply involved in the design process itself.

There is a symmetry to all this, a racing into the future in order to regain some of the splendors of the past—only this time not just for the select few. It seems likely that in the not-to-distant future customers may just find themselves in the role of the Honorable Evelyn Henry Ellis, M.P., designing a car to fit their own unique desires and seeing it delivered in a matter of weeks. In T.S. Eliot's words, "And the end of all our exploring/ Will be to arrive where we started/ And know the place for the first time."

# 6

## The Machinery of Change

The ultimate goal of manufacturing in the virtual corporation is to provide high-quality products instantaneously in response to demand. This concept, however, violates many of the truths that traditional manufacturing managers have held sacred. For years the manufacturing floor was dominated by dogmatism on the part of managers and inflexibility on labor's side. It began with Frederick Winslow Taylor's stopwatch and the notion of the "one best way" and quickly deteriorated into a deep hostility between labor and management that survives in many U.S. industries to this day.

Labor responded to the deskilled, routinized, *Modern Times* world of the interchangeable worker with defensive demands for rigid and inflexible work rules that made responsiveness to change almost impossible. At the same time, Fordism and mass production saw the construction of factories set up and tooled to build only a limited variety of products at very low cost. With these inflexible factories it necessarily followed that marketing should sell what the factory could economically produce. Manufacturing managers with little confidence in workers believed the only way to achieve productivity was to drive for high output using performance standards as the sole motivating tool. The accepted way to meet quality goals was to inspect after the product had been produced and rework defective production. Not surprisingly, over time management learned to assume that low cost came at the expense of both quality and flexibility.

Companies also began to accept poor quality and poor performance of suppliers as a necessary cost of doing business—a problem to be endured, not corrected. And, since suppliers shipped poor-quality products and did not deliver when they said they would, manufacturers learned to build large inventories of raw materials and components. Since capital equipment broke down frequently, companies planned for failure and learned to order redundant equipment. A "just in case" manufacturing philosophy took up residence at the heart of American industry. It was now OK to have too much inventory and banks of idle machinery.

When, after first experiencing the superiority of Japanese products during the 1970s, customers began to demand higher-quality U.S. products, manufacturers most often responded by adding more and more inspection steps to the just-in-case model. They created so-called inspection-oriented plants in which the cost of finding and reworking products was as much as 50 percent of production costs—hardly competitive with the low-cost Japanese counterparts.[1]

Faced with such poor records, in recent years the accepted dogmas have come under close scrutiny and the conventional wisdom has changed. As researcher Earl Hall has written: "The decade of the 1990s promises to provide a wealth of information and experience for establishing the modes of competitive manufacturing in the 21st century. Indeed, much of the restructuring of the last quarter of the 20th century, and industry's response to that restructuring, is prologue to the next century."[2]

Much of this change has come from unlearning the rules of the past. For example, modern companies have learned to strive for nearly perfect quality. A case in point, Motorola has set quality goals for its products at just three defects per million parts. As U.S. companies have studied the successes of lean production in Japan, they have come to realize that teaming up with suppliers can lead to the predictable delivery of reliable components, and that by working closely with capital equipment producers they can come to expect dependable machinery. And when this occurs, just-in-case becomes an anachronism, an embarrassing memory.

The differences between the just-in-case factory of yesterday and the lean production facilities of today are striking. When researchers Womack, Jones, and Roos visited the Toyota-Takaoka facility right after touring GM's Framingham, Massachusetts, plant they were stunned by the difference. Whereas the GM plant's wide aisles were crammed with indirect workers ("workers on their way to relieve a fellow employee, machine repairers en route to troubleshoot a problem, housekeepers, inventory runners") adding no value to the actual product, the Toyota plant's narrow aisles were almost deserted. "Practically every worker in sight was actually adding value to the car."[3]

At the individual workstations, the difference was equally obvious. In Framingham, weeks' worth of inventory was piled up, with defective parts "unceremoniously chucked in trash cans." At Takaoka, no worker had more than one hour's worth of inventory at his workstation—and when a defective part was discovered it was immediately tagged and sent to a quality control area for replacement.

Although only senior managers could halt the line at Framingham, the line regularly stopped anyway—thanks to machinery breakdowns and parts shortages. At any given moment, some workers were buried in work while others waited with nothing to do. The work area was filled with defective cars; the paint area, backed up with numerous car bodies awaiting work; and outside, railway cars sat filled for days with parts awaiting unloading into the giant parts warehouses.

By comparison, at Toyota-Takaoka there were no bottlenecks, no warehouses (when asked how many *days* of inventory were at the plant, the Toyota executive thought the word *minutes* had been mistranslated), and, although any worker could stop the line at any time, almost no work stoppages. As for defective cars awaiting rework or repair—there were none.

The reader will probably not be surprised to learn that the visitors found the Toyota workers to be purposeful and self-motivated, even though their work pace was quicker. As for the workers at Framingham, the only word the researchers found appropriate was "dispirited."

As companies began to wean themselves from a just-in-case mentality, they were also discovering that responsiveness and flexibility did not have to be traded off against low cost. As Stalk and Hout have pointed out: "Demanding executives at aggressive companies are altering their measures of performance from competitive costs and quality to competitive costs, quality, *and* responsiveness."[4] A reduction in the time it takes from design to tooling, combined with the flexibility to make short product runs with minimum plant downtime, results in short product cycle times. And that in turn gives the manufacturer the ability to more quickly respond to changing customer interests.

A growing number of examples of how cost, quality, and responsiveness are compatible can be found throughout U.S. industry, from small steel mills to application-specific integrated circuits manufacturers to clothing makers. So numerous have these cases become that not even the most hidebound industry executive can argue the conventional wisdom with much conviction.

Ford Motor Company, for example, has invested in a new flexible manufacturing plant in Romeo, Michigan, that will break with the traditional pattern at U.S. automobile engine factories of producing a half-million copies of a single engine type each year. According to *Business Week:*

> Ford thinks it has a better idea. First, design V-8 and V-6 engines around a basic building block—in this case, a combustion chamber designed for maximum fuel economy. Then, equip factories with machinery flexible enough to build several different models. Ford now plans as many as six new V-8s and V-6s in the 1990s—everything from cast-iron workhorses to high-performance aluminum thoroughbreds.
>
> At Romeo, flexible manufacturing equipment and modular design will permit production of more than a dozen engine sizes and configurations on one line. The engines will share about 350 parts. That will give Ford unprecedented freedom to match the plant's 500,000 engine capacity with customer demand.[5]

General Electric, often lauded for its early jump into flexible manufacturing, proved the power of this system a few years back at its locomotive works in Erie, Pennsylvania. These works were among GE's first successful attempts at creating responsive manufacturing. The test came when the locomotive business began to dry up. Could GE adapt to a changing market? As *Industry Week* later reported, "the line was switched to motor frames without missing a beat."[6]

Stephen Cohen and John Zysman have called this kind of manufacturing responsiveness static flexibility, which they define as "the ability of a firm to adjust its operations at any moment to the shifting conditions of the market—to the rise and fall of demand or the changes in the mix of products the market is asking for." They argue that this flexibility can be technological (the use of new programmable machine tools), political (the worker buy-out of a steel plant in Weirton, West Virginia, that results in lower wages in exchange for the workers' stake in the firm), or organizational (the reduction in the number of job categories at the GM-Toyota NUMMI plant from 183 at other GM plants to just 4). "Static flexibility decreases the risk that the firm won't be able to adapt to changes in the number and types of goods demanded in the market; it increases the ability to adapt to changed conditions."[7]

The idea of manufacturing responsiveness is tightly bound with that of total cycle time, the interval between when the market desires a product and when a company answers that need—in Stanley Davis's terms, "from conception to consumption." Davis, like many observers of industry, believes that anything more than marginal (10–20 percent) reductions in total cycle time are possible only with revolutionary change—"reconceptualizing the production, distribution, and/or delivery processes themselves."[8] These changes must take place because of shrinking product life cycles. In order to compete effectively, companies are being driven to flood the market with new products that take advantage of new technologies as soon as they are available.

The idea of time as a competitive advantage is not new, but it has always been an important consideration in business. As Alfred Chandler has written, many of the advances of the Industrial Revolution in America that have been credited to the economies of scale and distribution were not those of size but of speed. Chandler suggests:

> They did not come from building larger stores; they came from increasing stock turn. . . . It was not the size of the manufacturing establishment in terms of numbers of workers and the amount and value of productive equipment but the velocity and throughput and the resulting increase in volume that permitted economies and lowered costs and increased output per worker and per machine.[9]

What has changed is that time reduction has become a conscious program in most forward-looking companies, a component like any other that can be improved, modified, and even eliminated through intelligent planning and the judicious use of technology. The Japanese discovered this, with stunning results. In the automotive industry, they coupled the adaptability of lean production with short design cycles to blanket the market with new products. One very visible result is that the average age of Japanese car models is now about two years—compared with five years for North American and European models.[10]

None of this has been lost on companies in competitive high-technology markets. They experience it every day. In early 1991 a survey by Technologic Partners and Ernst & Young found that for technology company executives,

> the main cause for alarm . . . lies in expectations for further acceleration in product life cycles, which are already turning out new products faster than most customers can absorb them. . . . Where more than half of the systems companies saw product life spans of 36 months or longer five years ago, most now expect their products to survive for less than three years. And the overwhelming majority expect average product life span five years from now to be less than 18 months. Software companies don't go that far, but the same trend is at work.[11]

Recognition of this has led to a focus on time, or more precisely on its reduction, as a key factor in business success. Say Stalk and Hout, who pioneered this field of study, "Today's innovation is time-based competition."[12] And as former electronics industry executive Philip R. Thomas has written, "It is important to recognize that the big do not out-perform the little; the fast most frequently out-perform the slow."[13]

The only way companies can compete at this kind of pace and still turn an adequate profit is for them to race down the path toward becoming virtual corporations. Some firms have already begun the process and their examples serve as models to industry of what can be accomplished.

One much-reported story is that of General Electric's circuit breaker business. Challenged by Siemens in that market, GE concluded that what it needed was quicker response times to customers. "We had to speed up or die," said William Sheeran, general manager of GE.[14] To achieve that acceleration, the company consolidated all of its production into a single North Carolina plant, installed automated flexible manufacturing systems, and reduced its twenty-eight thousand different components by more than 90 percent—yet it still left its customers with more than forty thousand different product choices.[15]

The impact of these changes was dramatic. As *Fortune* reported, the time it took GE to fill orders fell from three weeks to three days. Costs fell 30 percent, while productivity jumped 20 percent, in the same year. The plant's return on investment rose to more than 20 percent and, in what was considered a tired market, GE actually gained market share.

Joseph Blackburn has reported on a similar effort by the Allen-Bradley Company to retain its market dominance in industrial controls and industrial automation. Installing a high-volume, flexible manufacturing facility that could take, manufacture, and ship orders in just twenty-four hours, the company saw its costs per unit fall to the lowest in the world, while at the same time its product offerings jumped from 125 to 600.[16]

Milwaukee's Badger Meter, a maker of flow meters for everything from private homes to city water mains, was an early convert to flexible manufacturing, installing its first two-machine work cell in 1986 to replace six stand-alone machines. The impact has been extraordinary. "Today," stated John J. Janik, vice president of operations, "all of the cast water-meter housing machines at Badger are automatically inspected and have to come out right the first time through. Each set of castings in a cluster is completed in six minutes. . . . With the six stand-alones, production of a group took twelve weeks, given all the fixturing and setups."[17] Just as important as this jump in performance has been Badger's improved customer response process. Said Janick: "Each morning the production manager simply keys into the program which and how many of the 110 different housings of brass, bronze, steel or aluminum he wants for the day based on orders received the day before."[18]

Du Pont, once enjoying a 90 percent market share with its rubbery plastic Kalrez, found itself losing market share to Japanese rivals. The company responded by shortening cycle times from seventy days to sixteen, reducing lead times from forty days to sixteen, and improving on-time deliveries from 70 percent to 100. Sales jumped.[19]

One of the most complete examples of a company revamping an operation around the virtual product model is that of Motorola's Bandit electronic pagers. Faced with heavy competition from Asia that had driven out all other U.S. pager makers, Motorola set about revising its production strategy. The objective, wrote Jagannath Dubashi in *FW* magazine, was "to make pagers of surpassing quality, customized to order with economies of scale." That is, virtual products.[20]

To do so, the Bandit team had to rethink everything about design and manufacture. The product was completely redesigned to be more buildable by the new generation of soft-automation, flexible, computer-controlled robots. A year and a half was spent benchmarking the manufacturing process at companies around the world (including numerous watchmakers, which had a side benefit of enabling the group to design wristwatch pagers a few years later). A new modular conveyor system from the United Kingdom was installed. With Wal-Mart and Benetton as their examples, the Bandit team installed a computer-based order entry system that would turn a salesman's order into activity on the manufacturing line in just twenty minutes.

In February 1988, less than eighteen months from the program's beginning, the Bandit line was under way. Order times were reduced from one month to minutes; manufacturing time, from five hours to three. And just seventy employees supervised twenty-seven robots as they assembled pagers with more than one hundred parts. So successful has been the program that Motorola has begun applying its lesson elsewhere in the firm.

Each of these examples, and there are scores more throughout industry, suggest the power of virtual product manufacturing. According to *The Economist,* the average factory takes an estimated ten to fifteen times too long to pass a product through.[21] Be it merely the linking together of a few milling machines into a flexible manufacturing system or the ultimate step of a fully computer-integrated factory, the challenge for companies is to gain control over the time component of their manufacturing—and then make the crucial next step to converting and consolidating that newfound advantage into a customer service.

## A Distant Trumpet

The fast responsiveness of manufacturing in the virtual corporation eliminates many of the errors caused by poor forecasting. In turn, short cycle times attenuate what business analysts call "the trumpet of doom," a plot of forecasting error versus time. What it implies is that the further a person must forecast into the future, the greater the possibility of error.

The trumpet of doom plagues the clothing and apparel industry, where clothes are ordered months before they are sold. The inability to know in spring what the consumer will want in the autumn continually leaves retailers with shelves stocked with the wrong product and a seemingly endless parade of seasonal sales. Retailers lacking virtual corporations as suppliers have come to accept their fate much as the manufacturers of yesterday dealt with similar problems with a just-in-case philosophy.

There is a better way, as Badger Meter has shown. By being able to set production each morning and reprogram the factory machinery in minutes, Badger spares itself the waste that comes from sudden cancellations or order revisions. Worker time is no longer wasted on production doomed to become scrap. This is similar to the situation at Benetton, with its custom dyeing to reduce waste.

The ability to respond better to customer needs enables the company to command premium prices. When one considers that the manufacturer is already enjoying increased profits from smaller inventories, less scrap from quicker production cycles, and lower costs, the chance to command higher prices comes as a welcome bonus.

This bonus is possible for two reasons. First, the product reaches the market sooner, increasing its perceived value to potential customers. Second, because of its better fit of features to the needs of those customers, as well as its greater quality and reliability, the product can carry a higher initial price. This is what Stalk and Hout have argued—that time-based products enjoy higher profits and faster growth in sales than their tardy competitors.[22]

That's at the beginning of the learning curve. In time, as competition increases, these products should still be better able to hold their prices, thanks to strong customer perception of quality. And when at last prices do begin to fall, these products have a deeper cushion of profitability to fall back upon. Finally, at the end of the product's life cycle, the company has a greater ability to predict where the market's changing needs will lie and a greater capacity to develop, build, and deliver a follow-up product to start the process all over again.

Manufacturing in the virtual corporation is based upon lean production (and all that it implies) and on a continuing flow of new technology that has made possible flexible and computer-integrated manufacturing and low-cost capital equipment. These, in turn, often make it practical to produce products in the distribution channel and at the customer's site.

Lean production is the name given to the Toyota production system by John Krafcik of the International Motor Vehicle Program (IMVP), a $5 million study of the future of the automotive industry by the Massachusetts Institute of Technology. The study concluded that

> lean production is "lean" because it uses less of everything compared with mass production—half the human effort in the factory, half the manufacturing space, half the investment in tools, half the engineering hours to develop a new product in half the time. Also it requires far less than half the needed inventory on site, results in fewer defects, and produces a greater and ever growing variety of products.[23]

Benjamin Coriat of the University of Paris has pointed out that lean production is much more an organizational innovation than a technological one.[24] It depends not so much on computers and automatic machinery as it does on worker skills, organization on the factory floor, and relationships between manufacturers, suppliers, and customers. It requires that suppliers deliver near-perfect product precisely when it is required and that manufacturers have production machines that are almost always available and working correctly whenever they are needed. (The former is one reason why Japanese manufacturers prefer to stick for decades with proven suppliers rather than risk even the most appealing new sources.)

The IMVP study contrasts lean production with craft and mass production in the following way:

> The craft producer uses highly skilled workers and simple but flexible tools to make exactly what the consumer asks for—one item at a time. Custom furniture, works of decorative art, and a few exotic sports cars provide current-day examples. . . . Goods produced by the craft method— as automobiles once were exclusively—cost too much for most of us to afford.
>
> The mass-producer uses narrowly skilled professionals to design

products made by unskilled or semiskilled workers tending expensive, single-purpose machines. These churn out standardized products in very high volume. Because the machinery costs so much and is so intolerant of disruption, the mass-producer adds many buffers—extra supplies, extra workers, and extra space—to assure smooth production. Because changing over to a new product costs even more, the mass-producer keeps standard designs in production for as long as possible. The result: the consumer gets lower costs but at the expense of variety and by means of work methods that most employees find boring and dispiriting.[25]

Lean production appears to do the impossible. It delivers the great product variety once associated only with craft production at costs that are often less than those associated with mass production. These benefits are provided along with products of extraordinary quality.

But lean production also has been widely misunderstood. It is not a solitary event but the continuous application of a process. A company becomes lean by making gradual improvements in each step of the manufacturing process. In fact, the creator of the Toyota production system, Taiichi Ohno, arrived at lean production as a result of his efforts to eliminate all waste.[26] His goal was to make the most efficient use of all of the resources—human, capital, and supplier—available to the company. In the process, he adopted total quality control methodologies and developed the just-in-time (JIT) system of manufacturing. Although technology has an important role to play in lean production—flexible tools, work cells, and computer-integrated manufacturing—it is not fundamental. What is fundamental is teamwork between production workers and management. General Motors found this out in its efforts to install the lean production methodologies at two of its factories.

Researcher Maryann Keller described the successful arrival of lean manufacturing at the GM-Toyota NUMMI plant in Fremont, California:

At the core of [lean manufacturing] were two premises—both new for GM. The first was that the average worker is motivated by the desire to do a job that enhances his sense of self-worth and that earns the respect of other workers. The second premise was that the worker is inspired by an employer who places value in the worker's input. Under the system in Japan, every worker is encouraged to think—to use brainpower to find ways of improving products and processes and eliminating wastes—and is rewarded for improvements.[27]

Compare this with Keller's description of life at GM's traditional plant in Van Nuys:

> The workers wanted to do their jobs well, wanted to be competitive, but all too often they were fighting against unbeatable odds to get the job done. Every problem became a confrontation; since there was a basic mistrust between labor and management, it was hard to establish a cooperative environment where problems could be solved. . . . Workers operated in a vacuum; they did the job they were told to do without relating it to the final outcome. . . . [It] was an environment where workers were not partners in the task of building cars. Often they did not know how their jobs related to the total picture. Not knowing, there was no incentive to strive for quality—what did quality even mean as it related to a bracket whose function you did not understand? Workers were held accountable through a system of intimidation: Do your job and your supervisor won't yell at you. That was a pretty thin incentive![28]

Lean manufacturing violates so many of the traditional beliefs about mass production that one might find it hard to believe that it will work at all. Ohno, for example, believed that the key to efficient production was not long runs of identical products but rather short runs of a wide variety of products. In order to accomplish such short runs he had to discover ways to eliminate the waste associated with long set-up times for production equipment in the manufacturing areas.

Some of the tricks that Ohno figured out involved working directly with tool and equipment manufacturers to make the process easier and storing the tools on the production floor rather than in tool cribs so they would be instantly accessible to the worker. In order to make such a system work effectively, Ohno also had to train his workers to understand how to perform the changeovers. He also had to trust them with tools that manufacturers traditionally store in secure areas for fear they'll be lost, damaged, or stolen.[29]

The results achieved by Ohno were astounding. For example, between the late 1940s and late 1960s, the time it took to change a tool dropped from as long as three hours to just three minutes—a difference of nearly two orders of magnitude that provided a clue to the dramatic innovations to come.[30] Once it was possible to reduce set-up times to the point that they were insignificant, it became practical to vary production almost at will and dramatically reduce in-process inventories. Workers could produce the parts needed by assemblers for the next hour's production and then change their set-ups to be ready for the production of a different product. (This is why researchers found less than one hour's inventory at any workstation at Toyota's Takaoka factory.)

However, if the worker were going to produce just what was needed by the person he or she was supplying, or if a supplier were going to supply parts just when they were needed, an efficient way had to be found to coordinate everyone's efforts. In 1947, when Ohno began his work on the Toyota production system, the Japanese did not have vast amounts of inexpensive computer power at their disposal. But even if they had, it is not obvious that a computerized system would have been nearly as efficient as the way Ohno discovered for doing the job.

## Just-in-Time

To meet the challenge of delivering parts when they were needed, Ohno developed the *kanban* system, which has since been adopted in many different forms and has come to be known as just-in-time, or stockless, production. *Kanban* has become synonymous with Japanese industry, so it is shocking to learn where it came from. As Ohno tells it:

> Kanban [is] an idea I got from American supermarkets.... Following World War II, American products flowed into Japan—chewing gum and Coca-Cola, even the jeep. The first U.S.-style supermarket appeared in the mid-1950s. And, as more and more Japanese people visited the United States, they saw the intimate relationship between the supermarket and the style of daily life in America. Consequently, this type of store became the rage in Japan due to Japanese curiosity and fondness for imitations.
>
> In 1956, I toured U.S. production plants at General Motors, Ford and other machinery companies. But my strongest impression was the extent

of the supermarkets' prevalence in America. The reason for this was that by the late 1940s, at Toyota's machine shop that I managed, we were already studying the U.S. supermarket and applying its methods to our work. . . . We made a connection between supermarkets and the just-in-time system. . . .

A supermarket is where a customer can get (1) what is needed, (2) at the time needed, (3) in the amount needed. . . . From the supermarket we got the idea of viewing the earlier processes in the production line as a kind of store. The later process (customer) goes to the earlier process (supermarket) to acquire the required parts (commodities) at the time and in the quantity needed. The earlier process immediately produces the quantity just taken (restocking the shelves). We hoped that this would help us approach our just-in-time goal and, in 1953, we actually applied the system in our machine shop at the main plant.[31]

Just-in-time systems have been defined as those that "produce and deliver finished goods just in time to be sold, subassemblies just in time to be assembled into finished goods, fabricated parts just in time to go into subassemblies, and purchased materials just in time to be transformed into fabricated parts."[32] Or, more simply, "the idea of producing the necessary units in the necessary quantities at the necessary time."[33] (The term is sometimes used in ironic juxtaposition with just-in-case, the mass production philosophy of keeping massive inventories on hand to cover any eventuality.)

The *kanban* system as conceived by Ohno was brilliant in its simplicity. The Japanese word *kanban* actually means "visible record." Basically, each lot of parts had a *kanban* card attached to it. When the parts were passed down the assembly process, the card was attached to them. When more of the part type was required, the card was passed upstream, the worker set up the machine to produce the part, and production commenced. The same system was used with suppliers as well. Most of them were located in close proximity to the Toyota plant, so *kanban* cards could be easily delivered.

When suppliers produce products of extremely high quality and deliver in exact quantity in response to *kanban* requests, much of the cost and overhead associated with mass production vanishes. There is no need for incoming inspection, raw material inventories, or sophisticated billing, ordering, shipping, and receiving procedures. The *kanban* card becomes the purchase order and a copy of it (or the actual card itself) can be sent to the accounting department and can serve as the bill from the supplier. In the process, much of the manufacturing overhead vanishes, along with accounting complexity. Production becomes lean not only on the factory floor but in the factory office space as well.

One of the authors of this book, when he was responsible for a factory, remembers reviewing a six-inch-thick stack of paperwork associated with the payment for the purchase of one thousand floppy disks. As he dug into the pile, he asked his controller why it was so difficult to figure out how much money was owed. The answer was simple: the quantities shipped were different than the quantities ordered. Say that fifty units were supposed to arrive on a particular day—sometimes fewer arrived and at other times more arrived. Seldom did the right quantity arrive at the right time. Of course, a large percentage of what was shipped had to be returned as faulty—and that meant even more paperwork associated with returning the material and tracking its replacement. Hours of work were expended on what would have taken minutes with a *kanban* system and reliable supplier. On top of that, the company had to maintain large amounts of just-in-case inventory.

Of all the new manufacturing processes, just-in-time is probably the most pervasive. In 1985 *Purchasing* magazine polled four hundred of its readers and determined that 20 percent had a JIT program in place and 10 percent were planning one. Two years later, with one thousand respondents, the figure had jumped to 72 percent who either had a program in place or were planning one; and more than 75 percent of those who implemented JIT methods reported improvements in quality and productivity.[34]

One reason for this jump has been recent recognition of the benefits of the JIT system for both high- and limited-volume production. Western companies, watching the Japanese use this methodology to produce limited selections of high-quality, low-priced products, often concluded that JIT was useful only in high-volume, long-production-run manufacturing. The firms decided that the greater product variety of job-lot production often found in Western factories would be the best defense against Japanese competitors armed with JIT. However, as we all now know, the truth was just the opposite. Writes Richard J. Schonberger:

Growth spawns variety. The company that has gotten rich making "basic black" will then build a plant to make some other model or product. In time, the company ends up with a full line of models, all produced and sold in volume and manufactured more or less repetitively—the present situation at Nissan, Sony, Canon, and other top Japanese companies. The Western job-lot competitor no longer has even the advantage of more product variations.[35]

Not that implementation of JIT will come easily. While U.S. industrial companies have jumped on it as a solution to many problems, their suppliers have proven far more resistant. Said Vaughn Beal, president of Harley-Davidson, one of the most famous JIT success stories, "The easy half of the job of implementing JIT is doing it inside. The tough one-half is doing it with suppliers." A study in the *Journal of Purchasing and Materials Management* found that 50 percent of automotive JIT users reported that poor supplier quality was a problem and that while 85 percent of these respondents had implemented JIT, only 39 percent of their first-tier suppliers had.[36]

One reason for supplier resistance to JIT is a fear that manufacturers will use it as a means for shifting blame up the chain and loading suppliers with inventory they once carried themselves. Smart companies have moved to assuage this concern by creating stronger supplier–manufacturer ties through shared information and even manufacturer training and investment in supplier JIT systems. Needless to say, it is precisely this sense of co-destiny, of shared participation in product creation, that is a precursor to virtual corporations (see chapter 7).

It is easy to see how, in the virtual corporation, JIT will play an important role. Affordable volume production of custom products will require strong control over inventories and production to maintain both competitive pricing and the ability to make nearly instantaneous product shifts to meet changing market demand. This can occur only if the manufacturer is not burdened with expensive and soon-to-be-obsolete inventory. The reduction of inventory in the system also eliminates the shock waves of frantic orders and cancellations that paralyze the supplier network as vendors struggle to control their inventories while being whipsawed by violent changes in customer demand.

For the just-in-time process to work, everything must be right. Machines must function when they are needed and not break down unexpectedly. Parts must be right when they arrive. If they are out of specification and cannot be used, the line will stop. One of the beauties of the just-in-time system is that it does stop whenever there is a problem. It therefore provides tremendous motivation to fix problems permanently so they will not occur again. By the same token, however, it would be a mistake to install a just-in-time system in a facility plagued with quality problems.

The virtual corporation is likely the most fertile environment for implementing JIT. A company built on the efficient and rapid gathering, processing, and distribution of information is one that can take JIT to new levels of precision. The ability to predict market changes takes just-in-time out of a reactive mode into an anticipatory one.

Nevertheless, it still seems likely that the dream of a perfect just-in-time production system will remain just that even in the virtual corporation. Pure JIT, devoid of inventories altogether, will be a rare event in the twenty-first century. Even the Japanese have found that in most situations, supporting pure JIT is simply too expensive.[37] Rather, each industry is likely to find an optimal mix of JIT and some inventories. This mix, which will change dynamically with the shifting operations of the virtual corporation, will depend upon such factors as type of product (or component), production facilities, organization, and location.[38]

## Total Quality Control

Hand in hand with just-in-time goes the race for perfect quality. When there is no margin for error, there can be no tolerance for performance short of perfection.

Emphasizing quality as a production methodology was an American idea, the life's work of such men as W. Edwards Deming and J. M. Juran. It was the Japanese, however, who put it to use to establish an almost unassailable reputation in the world marketplace. As Ohno has written: "Imitating America is not always bad. We have learned a lot from the U.S. automobile empire. America has generated wonderful production management techniques such as quality control and total quality control and industrial engineering methods. Japan imported those ideas and put them into practice."[39]

One important aspect of total quality control (TQC) is that it shifts major portions of the responsibility for quality from the staff operations of a firm to the manufacturing line. That is to say, obtaining ever-higher product quality is no longer solely the task of preproduction design or postproduction inspection but an inextricable part of the manufacturing process itself. As such, a product is not considered built unless it performs to specifications.

Juran himself has consistently refused to give a simple definition for quality, arguing that any such definition would be a trap. Instead, he opts for a multiplicity of meanings, of which two have critical importance: product performance/product satisfaction (he adds that "product" includes both goods and services) and freedom from deficiencies/product dissatisfaction. The first deals with characteristics of the product that lead users to become satisfied that they made the purchase—features such as ease of use, gas mileage, promptness of delivery, and so on. The second deals with those factors which make the customer unhappy with the purchase and lead to complaints, repairs, and returns. Juran stresses that the two meanings are not opposites of each another. Rather, the goal of product performance is to be better than competing products; the goal of freedom from deficiencies is to have perfect quality.[40]

The key to achieving total quality is getting everyone in the organization, from the president on down, involved in the process. All of the external suppliers must buy into the process as well. The goal is to eliminate all defects. In the process, employees, not just quality control specialists, are taught how to identify problems and analyze their source using a variety of diagnostic tools such as quality dispersion charts, defect frequency rates and trends, process control charts, and the so-called Ishikagawa fishbone diagram, which charts the causes and effects in a particular manufacturing process.[41]

Some organizations that have gone on to achieve high levels of quality have claimed that it has become such a part of the company culture that everyone knows it is his or her job to eliminate all defects. Management personnel in these advanced organizations further support and maintain these high-quality levels by setting examples and providing ongoing training programs.

The value of quality has come as a surprise to various American manufacturers and service suppliers. Many have been stunned by the realization that quality is, in fact, free. An early observer, Philip Crosby, estimated that the cost of poor quality could be as great as 20 percent and that, with processes that eliminate defects and waste, most of that money could be saved.[42] He went on to list his four commandments of product quality:

1. Definition: Quality is the performance to requirements.
2. System: The prevention of defects.
3. Performance standard: Zero defects.
4. Measurement: The price of nonconformance to perfect quality.[43]

The combined effects of lean production systems, JIT methods, and TQC processes are stunning. For example, in studying the impact of combined JIT/TQC programs, authors Charles O'Neal and Kate Bertrand found a wealth of success stories, including the following:

- Harley-Davidson reduced manufacturing cycle times for motorcycle frames from seventy-two days to just two, while increasing final product quality from 50 percent to 99 percent.
- Digital Equipment, at its Albuquerque computer workstation pilot line, reduced overall inventory from sixteen weeks to three, while reducing the defect rate from 17 percent to 3 percent.
- 3M's Columbia, Missouri, plant saw a 70-fold reduction in critical defects, appearance defects, and packaging problems.[44]

Other observers have found similarly spectacular case studies throughout American business. For example, *Business Week* reported that at yet another 3M plant (in St. Paul), a quality program in the production of two-sided industrial tape cut waste by 64 percent, reduced customer complaints by 90 percent, and increased production by 57 percent—all in two years. Another case, one of the most remarkable, involves Kodak's manufacture of plastic tips for blood-analyzer machines. In ten years the company cut defects from twenty-seven hundred per million to just two—and the only reason those two were bad was that someone had closed the package improperly.[45]

Other firms throughout the world have experienced similar results. In the early 1980s Philips Electronics, one of the world's most successful automobile headlamp makers, tried to sell its products to

125

Toyota—only to find the Japanese firm deeply disappointed with the quality of the lamps. Philips adapted Toyota's TQC measurement system and discovered that its own product quality was, according to one company executive, devastating. Said the executive, "Our old wisdom was not valid anymore." Philips persevered, however, pursuing the elusive Toyota goal, and improved packaging, manufacturing, and transportation. By 1990 the firm enjoyed considerable savings in wasted material, saw defect levels fall one hundred times, doubled its exports to Japan, and captured 60 percent of that market.[46]

The differences in the bottom line were easy to identify in the Womack, Jones, and Roos MIT study of the automotive industry. At the Toyota-Takaoka plant in 1986, it took sixteen hours to build a car in 4.8 square feet of workspace per vehicle per year, with .45 defects per car. At GM-Framingham, it took nearly thirty-one hours in 8.1 square feet, with 1.3 defects. Toyota was responding to the market more quickly, using half the labor hours, getting more output from every square foot of expensive plant space, using its capital equipment more efficiently, and still producing a higher-quality product with lower warranty costs because it had fewer problems.

If this didn't make the point, the GM-Toyota NUMMI plant did. In 1987, using American UAW workers, the plant built a car in nineteen hours in 7.0 square feet and, most important, matched Takaoka with .45 defects per car. Inventories, which averaged two weeks' worth in Framingham, fell to just two days' worth at NUMMI—still short of the two hours' worth at Takaoka but certainly a decided improvement, which cut the need for millions in working capital.[47]

The Toyota production system was transferable to U.S. suppliers as well. The *Wall Street Journal* reported on Bumper Works, a small Danville, Illinois, manufacturer of pickup truck bumpers that Toyota targeted as a test case for its system in the United States. Bumper Works, which already had an excellent reputation for quality, still managed to increase productivity by 60 percent and reduce defects by 80 percent.[48]

A warning must be issued, however: total quality control, lean production, and just-in-time processes are so attractive that there is a tendency on the part of some results-oriented American manufacturers to go for broke. In the process, they have moved too fast and reached too far—and the results have often been disastrous.

General Motors tried to cure its productivity, worker, and quality problems with automation. It ended up with a technological nightmare. The GM Hamtramck plant, opened just outside Detroit in 1985, was designed to be the pilot for the company's Saturn line as well as a showcase for GM's new commitment to technology. According to Maryann Keller, the plant had 260 robots for welding, assembling, and painting cars; fifty automated guided vehicles to serve the assembly line; and "a battery of cameras and computers that used laser beams to inspect and control the manufacturing process."

But, instead of a showcase, Hamtramck became a nightmare of technology gone berserk. The stories of robot breakdowns and miscues read like a 1950s B movie that might have been titled *Robots from Hell*. . . .

- Robots designed to spray-paint cars were painting each other instead.
- A robot designed to install windshields was found systematically smashing them.
- Factory lines were halted for hours while technicians scrambled to debug the software.
- Robots went haywire and smashed into cars, demolishing both the vehicle and the robot.
- Computer systems sent erroneous instructions, leading to body parts being installed on the wrong cars.[49]

Meanwhile, as this debacle was taking place at Hamtramck, a research team from MIT found that a Ford plant producing comparable cars using far less technology was showing the same productivity. And as if that weren't enough, in 1987 Toyota opened a plant in Kentucky that used little robotics at all—and quickly had twice the production rate of Hamtramck.[50]

The GM story is unfortunately not unique. Americans have tended to focus on short-term results and posting the best possible numbers on the next quarterly report. The Japanese, taking a longer view, have believed the best way to achieve results is to perfect the processes upon which they are based. They have also approached the process of improvement in a very conservative way, becoming great believers in taking small, incremental steps and pursuing the goal relentlessly over extended periods of time. The Japanese have a word for this approach: *kaizen.*

## Kaizen

Author Masaaki Imai calls *kaizen* "the single most important concept in Japanese management—the key to Japanese competitive success." His definition:

> KAIZEN means *ongoing* improvement involving *everyone*—top management, managers, and workers. In Japan, many systems have been developed to make management and workers KAIZEN-conscious.
>
> KAIZEN is everybody's business. The KAIZEN concept is crucial to understanding the differences between Japanese and Western approaches to management. If asked to name the most important difference between Japanese and Western management concepts, I would unhesitatingly say, "Japanese KAIZEN and its process-oriented way of thinking versus the West's innovation- and results-oriented thinking.". . .
>
> In business, the concept of KAIZEN is so deeply ingrained in the minds of both managers and workers that they often *do not even realize* that they are thinking KAIZEN.[51]

*Kaizen* can also be seen as Cohen and Zysman's second form of manufacturing responsiveness, dynamic flexibility, which they define as "the ability to increase productivity steadily through improvements in production processes and innovation in product."[52]

For Imai, *kaizen* is the philosophical underpinning not only of all Japanese business methods but also of the society itself. Specifically in manufacturing, he suggests that it is a customer-driven process in which it is assumed that all activities should eventually lead to increased customer satisfaction. This is accomplished by a different kind of management thinking that rewards people's "process-oriented efforts for improvement" and not just results alone.

It's not hard to see the relationship between quality control and *kaizen*—in most applications, *kaizen,* as a system of communicating ideas for improvement up and down the corporation hierarchy, is the most efficient vehicle yet found for improving quality. *Kaizen,* with its different system of motivation and reward, is vital to the success of quality control circles. It is this process that enables some Japanese firms to now look beyond quality that is taken for granted (*atarimae hinshitsu*) to the extraordinary notion of quality that fascinates (*miryokuteki hinshitsu*) and the confluence of technology and aesthetics that marks the virtual corporation.[53]

*Kaizen* is one of the few well-developed business improvement philosophies. In use, it shifts the corporate orientation away from the bottom line (results orientation) and the fitful starts and stops that come from dependence on product breakthroughs (innovation orientation) toward a continuous, gradual slope of improvement in products, quality, and cycle times. As Cohen and Zysman see it, "American firms tend to produce product innovations periodically, moving from one plateau of best practice to another. The Japanese, studies suggest, move through continuous and interactive product innovation, steadily improving the production process." One result of this difference in philosophy, they add, is that the "Japanese system, with its greater dynamic flexibility, has achieved greater productivity gains over the last few years than the more rigid American one."[54]

One of the greatest strengths of *kaizen* is the speed with which it can incorporate the latest technological advances—an important factor when one considers that many studies of economic growth since World War II have concluded that technology, more than capital or labor, has been the dynamic force behind economic development and improvements in productivity. One reason why a *kaizen*-driven company can do this so efficiently is that its entire work force is oriented toward locating new ideas and swiftly and effectively putting them to work. In essence, every employee becomes a management consultant.

Different companies have different techniques for implementing *kaizen*. One of the most interesting ones comes, not surprisingly, from Taiichi Ohno. It is called the "five whys" and it operates in a way its name suggests. When a production problem is encountered at Toyota, one is expected to ask, five consecutive times, why it happened, each time with a greater level of precision until eventually the symptoms are overcome and the root cause emerges.

Suppose a machine stopped functioning:

1. *Why* did the machine stop?
   There was an overload and the fuse blew.
2. *Why* was there an overload?
   The bearing was not sufficiently lubricated.
3. *Why* was it not lubricated sufficiently?

The lubrication pump was not pumping sufficiently.

4. *Why* was it not pumping sufficiently?

The shaft of the pump was worn and rattling.

5. *Why* was the shaft worn out?

There was no strainer attached and metal scrap got in.

Repeating *why* five times, like this, can help uncover the root problem and correct it. If this procedure were not carried through, one might simply replace the fuse or the pump shaft. In that case, the problem would recur within a few months.[55]

To illustrate the power of *kaizen* in action, Imai offers the example of Nissan's Tochigi plant, which in the decade after the 1973 introduction of its first welding robot increased its automation rate to 93 percent and its robotization rate for welding work to 60 percent—while at the same time reducing standard work time by 60 percent and improving efficiency by as much as 20 percent. This remarkable metamorphosis occurred, according to a Nissan executive, through a series of *kaizen* campaigns that searched for plant improvements in increments as short as six-tenths of a second. Imai quotes one staff engineer as being told by his boss on the first day on the job: "There will be no progress if you keep on doing the job exactly the same way for six months."[56]

Through the use of *kaizen* the Japanese have chipped away at their manufacturing problems. They have improved quality here, reduced a set-up time there; replaced a poor supplier with a better one; worked with capital equipment suppliers to get them to produce equipment that was more reliable, easier to maintain, and faster to set up; trained their components suppliers to do a better job; and in the end improved quality and JIT performance to the point where lean production became possible. Needless to say, by definition, none of this happened overnight. Taiichi Ohno started his pursuit of lean manufacturing in the late 1940s and didn't feel that he had the process working reasonably well until the early 1970s.

This suggests that companies cannot begin producing virtual products overnight in their factories through some sort of massive reorganization or design project or capital equipment investment. Quite the contrary: rebuilding a factory for a virtual corporation requires replacing almost every brick in the old plant. Do that too quickly and the structure will collapse. The only practical way is through *kaizen.*

An interesting story illustrating this point is told by Masaaki Imai:

> Back in the 1950s, I was working in the Japan Productivity Center in Washington, D.C. My job mainly consisted of escorting groups of Japanese businessmen who were visiting American companies to study "the secret of American industrial productivity."
>
> Toshiro Yamada, now Professor Emeritus of the Faculty of Engineering at Kyoto University, was a member of one such study team visiting the United States to study the industrial-vehicle industry. Recently, the members of his team gathered to celebrate the silver anniversary of their trip.
>
> At the banquet table, Yamada said he had recently been back to the United States in a "sentimental journey" to some of the plants he had visited, among them the Rouge River steelworks in Dearborn, Michigan. Shaking his head in disbelief, he said, "You know, the plant was exactly the same as it had been 25 years ago."[57]

The lesson of Imai's story is that while a heavy reliance upon masterstrokes has been the bane of U.S. manufacturing, so has inertia, an acceptance of the status quo—*kaizen* and the route to becoming a virtual corporation lie in between. No doubt in the rapidly changing business environment to come some companies will speed ahead too fast with reckless disregard for what is practical. Others will remain wedded to the past. But the ones who take seriously the message of just-in-time, total quality control, and *kaizen* will effectively confront the challenge.

## The Illusion of CIM

The revolutions in manufacturing and information processing have occurred simultaneously. So, it is not surprising that there has been an ongoing interest, and sometimes a pressure, to put the latest electronic miracle to work on the plant floor.

In some instances, this process has worked beautifully. There are a number of examples of numerical control machines, basic robotics, and a host of other technology-driven tools and systems having had a very positive effect upon manufacturing.

With the tools proving themselves so useful, there has been a subsequent drive to integrate them into clusters to assume ever-greater production responsibility. Each of the first few levels of this integration—such as can be seen in Remington's computer-controlled machining of guns, Ford's basic building-block production of a variety of engine models, and GE's speedy manufacture of circuit breakers—is usually described as a flexible manufacturing system (FMS). More specifically, FMS is defined as a "computer controlled manufacturing system using semi-independent numerically controlled (NC) machines linked together by means of a material handling network."[58]

FMS was the next logical step in a process that had begun with very high-volume, low-complexity transfer lines containing dedicated machines and had progressed, with the use of computers and numerically controlled machines, to the automation of more sophisticated processes. FMS was targeted at bringing cost-effectiveness to what was described as midvolume/midvariety applications.[59]

In application, FMS places a series of numerically controlled and other types of machines under the management of a computer system that controls the sequential or random production of a family of parts. The machines are physically linked together by a material handling system. The computer's role is "to continuously monitor the activities of the equipment and provide supervisory and engineering reports. Simulation can be used effectively to predict the behavior of system components and, therefore, provide for appropriate corrective behavior when needed."[60]

Implemented properly, FMS can be a potent tool. It lends itself neatly to all the other components of manufacturing virtual products—JIT, TQC, *kaizen*—and in the process helps the manufacturer achieve many of the goals of the virtual corporation, including shorter cycle times, a smaller and smarter labor force, smaller lot sizes, and improved short-run responsiveness and long-term adaptability.

Unfortunately, FMS is often not properly implemented. Furthermore, many U.S. firms, convinced that computerization is the universal key to manufacturing success, have leap-frogged basic FMS into what is called manufacturing cell, or, more popularly, computer-integrated manufacturing (CIM). [Although unfortunate, such a move is certainly in keeping with Imai's claim that, whereas *kaizen* is people oriented, American-style innovation is technology and money oriented.] CIM takes the flexibility of FMS to another level of integration, one that perpetually finds the optimum match between labor and automation to achieve the maximum value of both.[61]

CIM is wonderfully appealing in theory, but often for the wrong reasons (lights-out factories, no more labor troubles, a quick edge on the competition). That's why CIM, touted as the big manufacturing breakthrough of the age, has turned into a disaster at many companies. And the reason is simple: to implement computer-integrated manufacturing requires the kind of sweeping organizational changes that many companies investing in CIM have yet to make.

Clues to this costly fiasco could already be seen with FMS. A survey by Professor Jaikumar in 1986 found that American flexible manufacturing systems exhibited an "astonishing" lack of flexibility, sometimes performing worse than the equipment they were designed to replace.[62] One example was Ford's St. Louis minivan plant, which had to go with a less appealing and heavier floor plan for one of its vehicles because the computer-controlled line was too inflexible to handle a better design.[63]

Poorly implemented FMS systems result in a poor use of machine tools, increased lead times, increased in-process inventories, and inefficient use of floor space.[64] And this has been found in only medium-complexity manufacturing; CIM, designed for high complexity, merely took the failure to a higher level. The most notorious case, once again, is General Motors, with its estimated $70 billion investment during the 1980s in computers and robotics to little apparent positive effect.

Tellingly, the companies that did succeed with CIM were often those, such as IBM, that needed it least. For those firms, CIM was merely the next step in a program of continuous improvement in efficiency that began with first understanding the flow of parts and information through the factory and then bringing on information-intensive tooling and reporting systems. Wrote researchers at Touche Ross: "There is a necessary progression from quality to dependability to cost to flexibility."[65] And in the words of an Association for Manufacturing Excellence report on flexibility,

> the strategy of flexibility cannot take root unless the environment for it is right. There are a number of assumptions about the conditions necessary for flexible organizations. One is that, for practical purposes, companies

are regarded as the network of people who compose them. Flexibility cannot be purchased by acquisitions or merger. Flexibility is something that people *do*. Organizations of people may be combined or dissociated, but flexibility itself is not attained through things, financial or physical.[66]

Searching through the rubble of failed or failing CIM programs and then comparing what they found with successful programs, managers and academic researchers have made some important discoveries. The most important of these was that the term *computer-integrated manufacturing* itself was a dangerous misnomer. It was not computers that were being integrated into the manufacturing environment, but people—hence the growing use of the replacement phrase *human-integrated manufacturing (HIM)*. According to Earl Hall, HIM, unlike CIM, recognizes that it is crucial to determine the most efficient manufacturing practices before automating them, that software and communications network require greater resources for development and maintenance than was thought, and, most of all, that computer-based manufacturing systems must "be designed for the total manufacturing environment, including the important human interface, and not as only productivity enhancements replacing the human being."[67]

Similar sentiments were heard elsewhere. "The early focus [mostly technological] has been broadened to include cultural [people-oriented] issues and an emphasis on communications," says Peter C. Graham, CIM market development manager for Digital Equipment. Adds Jack Conaway, DEC's manager of CIM strategic programs: "In CIM, doing the right things in the acquisition or implementation step must proceed from an understanding that the content of CIM is 80 percent cultural and only 20 percent technological. And unless an organization structures itself for an integrated operating mode, the acquisition process will fall short of the success hoped for."[68]

What does this mean for daily business life? Obviously, HIM demands a better-trained work force. Wrote one observer: "If the human factor behind all this slick new technology is ignored or poorly managed, even the best physical and technical changes will fail miserably."[69]

American workers have traditionally resisted any form of automation, based on the not entirely inaccurate assumption that companies want to use it to displace them. The irony for both management and labor is that an effective HIM program usually increases not only the role of direct labor on the factory floor but also its influence throughout the organization. The so-called wisdom of the shop plays an important role in the virtual corporation.[70]

One person who has long understood the real role of computers in manufacturing as the facilitator of labor is the always prescient Taiichi Ohno. In his book *Toyota Production System*, Ohno makes these astonishing comments about the structure of a manufacturing firm:

> A business organization is like a human body. The human body contains autonomic nerves that work without regards to human wishes and motor nerves that react to human command to control muscles. The human body has an amazing structure and operation; the fine balance and precision with which body parts are accommodated in the overall design are even more marvellous. . . .
>
> At Toyota, we began to think about how to install an autonomic nervous system in our own rapidly growing business organization. In our production plant, an autonomic nerve means making judgements autonomously at the lowest possible level; for example, when to stop production, what sequence to follow in making parts, or when overtime is necessary to produce the required amount.
>
> These discussions can be made by factory workers themselves, without having to consult the production control or engineering departments that correspond to the brain in the human body. The plant should be a place where such judgements can be made by workers autonomously.[71]

Ohno continues the metaphor even further, suggesting that businesses succeed because of their reflexes—their ability to react to the slightest change—and that the larger a business, the better such reflexes it needs. He also adds that companies can stiffen and slow with age just as people do.

Finally, this latter-day Hobbes turns to the mind of his corporate leviathan. Throughout millennia, mankind has developed, Ohno suggests, an "agricultural" mind, an attitude about time and resources that has often been at odds with the industrial world. Now, some wish to go to a "computer" mind in one jump. That would be a mistake, he says, because we must first pass through the intermediate step of an "industrial" mind. "The industrial mind extracts knowledge from working

135

people, gives the knowledge to the machines working as extensions of the workers' hands and feet, and develops the production plan for the entire plant including outside cooperating firms."[72]

Ohno notes that American mass producers now use computers extensively. So too does Toyota, "but we try not to be pushed around by it. . . . We reject the dehumanization caused by computers and the way they can lead to higher costs."[73]

In the United States, where business imagery typically deals with organized sports or warfare, such an anthropomorphic view of the corporation would be inconceivable—until recently. Consider this quote from a well-known *Harvard Business Review* article:

> The manufacturers that thrive into the next generation will compete by bundling services with products, anticipating and responding to a truly comprehensive range of customer needs. Moreover, they will make the factory itself the hub of their efforts to get and hold customers—activities that will now be located in separate, often distant, parts of the organization. Production workers and factory managers will be able to forge and sustain new relationships with customers because they will be in direct and continuing contact with them. Manufacturing, in short, will become the cortex of the business. Today's flexible factories will become tomorrow's service factories.[74]

Considering these arguments, Coriat's statement that lean production was more an organizational invention than a technological one becomes all the more cogent. The factory of the future will be more humanistic (and humane) than many people think.

## The Factory on the Move

While technology will not solve all of the manufacturing problems facing the virtual corporation, it will certainly present it with a host of opportunities. One of these is an unprecedented degree of flexibility in where to locate production facilities. With instantaneous worldwide communications it is theoretically as easy to control a factory in Asia as it is to control one right next door. The ease with which production can be integrated around the world will provide companies with greater flexibility than ever in selecting plant locations. As Hall has written: "The 21st century manufacturing company will extend the concept of the integrated production line to include the wide area transportation and communication network which supports that production line."[75]

The virtual corporation, however, will abhor distance. If it can find a friendly environment close to customers, it will want to locate there. Being close will enable it to be more responsive. There will be more opportunities to deal face-to-face with customers, to understand their problems, and to design products that precisely meet their needs, and there will be less chance of something going wrong as the product moves from the factory to the customer. One likely result of this will be a repatriation to the United States of production operations moved overseas a generation ago.

Speed has also turned the world upside down. Flexible factories, in many cases, don't have to be as large as mass production ones in order to be cost-effective. The inflexible machines of the mass production factory had to manufacture massive numbers of units in order to justify their existence.

Thanks to the rapid pace of technology it is now possible to build sophisticated factories at very low cost. Factories will be located centrally in some cases, in the distribution channel in others, and at the customer's site in other instances. The location of the factory will be a function of the capital intensity of the process and the skill level required to effectively operate it. Thus, automobiles, steel, aluminum, paper, dynamic RAMS, synthetic fibers, and pharmaceutical products will still for the most part be made in remote central locations. The capital intensity of the processes, the skill to achieve satisfactory results, and the desirability of locating these facilities in certain geographic areas dictates this. But many other products and processes do not face such constraints.

This book began by noting a number of changes in manufacturing. Low cost and highly automated systems make it possible to process color film in one hour in low volume at numerous points around the country. Prescription glasses that used to take weeks to have made are now commonly available in one hour. The machinery to grind the lens is simple enough to use and inexpensive enough to manufacture, making it possible to place the factory in the distribution channel.

If the production process can be made simple and inexpensive enough, it becomes possible to move it to the customer's plant, even into the consumer's home. Camcorders have made it possible for just about anyone to produce his or her own movies. Individuals can measure their own cholesterol levels, blood sugar, and blood pressure, and women can often determine whether they are pregnant without ever getting a doctor or clinical laboratory involved.

If virtual products can be put into the customer's hands, why not virtual services? Of course they already exist. Companies and individuals using computers are already doing banking transactions from their homes and offices. Again, thousands of ATMs have put banks on every street corner; and airline reservation systems make reservations and print tickets easily and inexpensively at remote locations—making it possible to locate a travel agency almost anywhere. There are also the new forms of libraries. Vast amounts of information have been stored in data bases in remote computer systems, which can be searched to find references in millions of documents, thus rendering obsolete old card files and the dog-eared *Reader's Guide*s. In much the same way, lawyers can use Lexus to find legal precedents and authors can search through years of the *New York Times* clipping morgue from their homes using the Source.

Of course, though we become so accustomed to it as to forget, all of us are also living with electronic stores in our own homes. Flip on the TV and you can buy products from the Home Shopping Network. Open the mailbox and discover Fifth Avenue's best stores at your fingertips. And you can use Federal Express to get products delivered the next day. The suburban shopper can now get products from an electronic Tiffany's faster than he or she can from downtown stores, the latter being an onerous journey that often must be planned weeks ahead.

Manufacturing in the virtual corporation is highly responsive and mobile. As technology moves ahead, it will be possible to put more and more of it in the customer's hands, anywhere in the world.

The factory of the virtual corporation may seem too good to be true. Custom products produced faster and at lower prices than mass-produced ones? It would be a preposterous notion if we weren't already seeing it happen. For once, both the producer and the consumer win.

# 7

## Shared Dreams

I n the late 1970s, Xerox, facing copier costs as much as 50 percent greater than its Japanese competitors and watching its once dominant market share dwindle, set out to regain its competitiveness. To do so, the company knew that it had to make severe cuts in production costs, which in turn meant revising the very nature of its relationship with suppliers. In the words of James Sierk, group vice president for quality, "To get ten times better, you have to change the process."[1]

The first step was to slash away at the supplier lists. Xerox set a goal of reducing the number of suppliers from five thousand to five hundred, the maximum it could handle if it were going to work closely with them to manage their performance—Sierk admitted that they could never have trained five thousand suppliers. At the same time, the company announced a goal of just one thousand parts rejected per million on its assembly lines—compared with as many as twenty-five thousand rejects per million being delivered at the time by some suppliers.

Ultimately, Xerox cut back to just 325 suppliers, a number that in time stabilized at 400. With this manageable group, the company embarked on an intensive training program, including instruction in statistical process control, statistical quality control, just-in-time manufacturing, and total quality commitment.[2] Suppliers that didn't keep up were dropped, sometimes en masse.

The result was that between 1981 and 1984 net product cost was reduced by nearly 10 percent per year, and rejection of incoming materials by 93 percent. New product development time and cost were cut by 50 percent and production lead times fell from fifty-two weeks to just eighteen.[3] By 1989, the defect range was just three hundred per million, and the company's Business Products and Systems Group won the Malcolm Baldrige National Quality Award.[4]

Xerox had reduced its supplier base by more than 90 percent. Nine out of ten companies that had been doing business with Xerox at the beginning of the process were no longer receiving purchase orders. The remaining suppliers saw their businesses grow.

## Reshuffling the Deck

Xerox, once the most secretive of American companies, is a harbinger of the future cooperative business environment. So too are former antagonists IBM and Apple Computer, having reached a business agreement in 1991. And the General Motors supplier councils; Genecor, the joint venture of Genentech and Corning Glass Works; and SEMATECH and other industry or regional consortia. And let us not forget the investment of Ford and Volvo in Hertz; of Ford, Tenneco, and Kubota in Cummins Engine; and of Boeing and GE in United Airlines.[5]

Like all industrial transformations, the new business revolution is forcing a revision of traditional corporate arrangements toward what Harvard professor Benson Shapiro calls "the new intimacy." As the rapid gathering, manipulating, and sharing of information become a preeminent process and as company boundaries grow increasingly fluid and permeable, established notions of what is inside or outside a corporation become problematic, even irrelevant.

Also becoming irrelevant are the once obvious differences among suppliers, manufacturers, distributors, retailers, customers, even competitors. At any one time, an enterprise or an individual may play multiple roles. For example, the role of manufacturer might be moved up the chain and placed with the traditional supplier, as with semiconductor companies assuming much of the work once done by computer makers. Or the manufacturer's role may move downstream to distributors (value-added reselling of computer networks), retailers (photo processing), and even consumers (creating, from components, home computer and audio/video entertainment systems). By the same token, suppliers can bypass the putative manufacturer to become distributors, or they can obtain market information directly from customers, as does Burlington Mills.

Customers have also been enlisted into the campaign, helping to design their Japanese cars, performing diagnostics and simple repair work on their Xerox machines, and bypassing the retailer and distributor at warehouse stores and manufacturers' outlets. Why involve customers so deeply? Because often the customer is in a better position than is the supplier to design the products he or she wants. In the words of Derek Leebaert of Future Technology:

> In the future, there will be continuous anticipated feedback to the producer from the eventual customer. The "end" of a transaction will be defined far closer to the real end. A new degree of control, by the hours rather than by the months, will change our notion of a finished product. At present, making changes in a major industrial process is like turning a supertanker. Like Alexander Pope's spider "living along the line," all parts of the process will be inter-feeling.[6]

Some industry observers have also predicted a rise in dynamic multiventuring—multiple-player joint ventures, sometimes even between erstwhile competitors, in which teams from the different companies come together to work on a common technology or product goal. Certainly, the Apple/IBM (and Motorola) arrangement suggests such a possibility, as did the AT&T/Marubeni/Matsushita Safari computer project. But it remains to be seen whether this will develop into a broad trend, whether companies will be willing to move beyond arm's-length relationships to actually surrendering sovereignty.[7]

Needless to say, this widespread reconfiguration of business relationships is likely to create complementary internal reorganizations. There does not appear to be a universal template for such a corporate structure. On the contrary, the constant shifting of roles suggests that the virtual corporation may exist in a state of perpetual transformation. As BRIE research suggests, the future structure of successful firms may also involve becoming more aware of differences in national busi-

141

ness cultures—from Japanese megacorporations to the assemblages of small factories in northern Italy—and adapting to various forms of organized labor, management styles, laws, social mores, and even familial relations.

Yet, from another perspective, virtual corporations will be more stable and unchanging than their contemporary counterparts. The nature of the new business relationship will result in stronger and more enduring ties based on a mutual destiny, one shared by groups of both suppliers and customers.

Thus, the virtual corporation may appear amorphous and in perpetual flux, but it will be permanently nestled within a tight network of relationships. In studying the nature of that network, it is best to look beyond the old categorizations and instead evaluate the flow of information, goods, and services to suppliers and through the distribution channel to end users. As Michael Borrus of BRIE has said, "For the virtual corporation, the key is no longer ownership of processes, but control of results."[8]

A common future and mutual support will be the hallmarks of relationships between suppliers and customers, who will increasingly share the same fate. For either to succeed, they both will have to prosper. The spirit of buyer and seller relationships of the future is captured in the term *co-destiny*—each player's destiny will be joined with that of the other. And mutual dependence will characterize the relationships among virtual corporations.

Customers will be very dependent on suppliers, having invested very heavily in these relationships. They will have shared their business secrets with them, trained them according to their needs, and integrated them into their design processes. Not surprisingly, the excellent supplier will be viewed as irreplaceable.

Suppliers will view customers in an identical fashion. A good customer will be a precious asset—hundreds of hours of management time will have been spent to secure key relationships with it. Suppliers' businesses will be restructured to meet the needs of the customer base. Plants may even be relocated to better respond to customer needs. In effect, losing a good customer will be a business crisis.

Companies are increasingly going to structure their businesses around the market segments they serve. The products they design will be tailored to their respective markets. The service infrastructures they put in place will have to meet the needs of specific customer groups. For all but the most basic commodity products, companies will find themselves bound to market segments and the customers who belong to them.

## Supplying the Virtual Corporation

The implications for new types of interactions between suppliers and manufacturers has been a hot topic for discussion in recent years. For example, authors Michiel Leenders and David Blenkhorn have identified the need for customers to do a better job of getting suppliers involved in ensuring the success of their customers:

> Supply is part of the competitive edge and of continuing top management concern. . . . The greatest contribution potential for supply stems from early involvement in the development of new projects, products, services, and organizational strategies. . . .
>
> A good supplier lives up to the deal made with respect to quality, quantity, delivery, price, and service. A better supplier goes well beyond this minimum by taking the initiative to suggest ways and means whereby the customer can improve products, services, and processes. An exceptional supplier places the customers' needs first and is in tune with the long-range objectives and strategies of the customer. The exceptional supplier has the mission to make the customers prosper. Good, better, and exceptional suppliers are scarce. Therefore, supply management can be called the battle for good suppliers.[9]

What Leenders and Blenkhorn argue is that as competition becomes more global and product cycle times become shorter, the traditional arm's-length relationship between suppliers and buyers must end. Buyers will depend ever more upon their suppliers not only for fast and reliable delivery but also to play an even greater role in the design and manufacture of the finished product. Leenders and Blenkhorn advise companies that if they wish to get high levels of involvement from suppliers they have to sell them on the benefits of being an exceptional supplier. They call this selling process "reverse marketing."

Leenders and Blenkhorn aren't alone in calling for this change. Dr. Brian Joiner, president of Joiner Associates, a business consulting firm, has said, "There are many places where close relationships between customers and suppliers solve problems. Nowadays I think a long-time close relationship between customers and suppliers is critical to survival."[10]

Adds Professor Robert Spekman of the USC School of Business:

> It has become obvious to many manufacturers that their ability to become world-class competitors is based to a great degree on their ability to establish high levels of trust and cooperation with suppliers. . . . Those who subscribe to the partnership approach to vendor management argue that the gains far exceed the potential risks.[11]

For a virtual corporation to succeed, it must be so closely linked with its suppliers as to create a shared destiny. To achieve that, many of the established barriers between buyer and supplier must be removed. Ultimately, even the boundaries between the two will become indistinct. For example, it will not be unusual for a supplier to have offices on its buyer's premises or for the players to share trade secrets, to cross-license patents, provide one another with cost information, and include the other in long-range planning.

Suppliers, suggests Jim Morgan of Applied Materials, must also begin to assume more of the responsibility for managing the relationship—until now almost wholly the job of the buyer.[12] One area where such customer management is often necessary, as a BRIE study of auto parts suppliers in New York State found, is in demanding that different customers all ratify the same electronic data interchange systems. Otherwise, as the study found, the supplier ends up becoming less productive because it has to work with three or four computer systems at the same time.[13]

Building such bridges between manufacturers and their suppliers does not come easily. It isn't simply a matter of sitting down with the supplier and announcing that "we need to work together more closely." Rather, it requires unprecedented levels of trust and commitment in placing the fate of a company in the hands of people who aren't even employees. The nature of this heightened trust can be great enough to make even the most open-minded CEO wince. As reported in *Industry Week:*

The road to world-class supply chain management meanders through a series of cultural changes—to a new plateau of trust. To achieve true partnership, customers and suppliers must share information—on new product designs, internal business plans, and long-term strategy—that once would have been closely guarded.[14]

Suppliers are no more likely to accept this new openness than their customers. S. Charles Zeynel, director of quality at Union Carbide's Chemicals and Plastics Group in Danbury, Connecticut, got just such a reaction when he proposed supplier–customer joint teams: "[Managers] were concerned that if you [include] customers on a joint team the customers would learn all your secrets, your weak points, and problems."[15]

The demands of just-in-time delivery make it almost impossible for suppliers and buyers not to share both tactical and strategic information. But it is very difficult for companies to engage in information sharing unless they trust each other's motives. Joseph A. Bockerstette of Coopers & Lybrand has observed, "Many suppliers feel just-in-time is a way for Fortune 500 companies to dump on them."[16] As long as suppliers feel this way, it is very unlikely they will give their full support to such a program.

Nowhere will this be more painful than in the sharing of cost information. This violates one of the biggest taboos in all of business. It will require an act of trust far greater even than sharing proprietary technology. Yet in the new business environment this will be required of both supplier and manufacturer so that each may help the other reduce process steps, set realistic tolerances, and in any other way contribute to their mutual success.

## Making the Cut

As manufacturers begin scrutinizing their supplier lists and reducing them to a manageable size, they are likely to keep only those with the most reliable offerings and closest relationships. This narrowing of supplier lists will characterize business at least through the end of the century. As Joel Dreyfuss has written in *Fortune*, "Suppliers had better learn fast. Most large U.S. manufacturers are reducing their number of vendors in order to control quality."[17]

A survey by Grant Thornton of 250 midsize manufacturers found that 69 percent had imposed stringent quality improvements on their suppliers and that 76 percent had dropped at least one supplier as a result. Control Data's Cyber division, for example, dropped 650 of its 800 suppliers. Sun Microsystems not only cut its suppliers from 450 to 150 but gave 80 percent of its business to just 20 of the remaining firms.

The list goes on: Dell Computer demanded not only that its suppliers meet a specific certification requirement but that the suppliers of those suppliers do the same. Harris Electronics Systems Sector cut its supplier list from 2,500 to 270. Harley-Davidson went from 320 suppliers to 120. And supplier reductions aren't going on only in the United States. After tough audits of its suppliers, BMW dropped 100 that did not measure up—and regularly audits the rest.[18]

Unfortunately, the firms most likely to suffer from this selection process are the small suppliers. Many are insufficiently capitalized to make the requisite investments being demanded by their customers. For others, installing the flexible manufacturing system required to support the customer may represent too much of an operational leap from the current struggle of just getting products out the door. Still others may not have a sufficiently broad product line to be a valuable sole source. And, says Professor L. Joseph Thompson of Cornell, "Small companies are less likely than large companies to have made improvements in productivity. They're concerned about meeting the payroll and not about the longer term."

Large or small, suppliers will have to scramble to make the cut with their increasingly demanding customers. In some cases, this will lead to considerable pain as relationships of many decades end in lapsed contracts and even bankruptcies.[19]

For the big supplier this heightened competition may mean not only upgrading manufacturing and improving information processing but even cementing relationships through equity investment. In recent years there have been a number of cases in which major suppliers have bought debt and stock in key companies in order to guarantee future contracts. That's what Ford did when it put $1.2 billion into Hertz's $1.3 billion buyout—after all, Hertz, which began buying Fords in 1918, is that automobile company's largest customer. And when United Airlines' parent Allegis Corporation faced a takeover in 1987, Boeing helped by buying $700 million of Allegis preferred stock. Soon there-after, Allegis placed a $2.1 billion order with Boeing.[20]

Investment can even help suppliers buy into new business rela-tionships. "We'd been trying to do more business with Payless for years," John C. Nicholls, Jr., of Masco Corporation told *Business Week* in explaining why Masco and a sister firm had purchased $209 million of new Payless debt and preferred stock. Masco soon enjoyed a multi-year contract with Payless. "It's our expectation they will become an even bigger customer," said Nicholls.[21]

Small suppliers, without such fiscal resources, must look for other ways to assert their irreplaceability. One method is to simply outper-form larger rivals in terms of quality, flexibility, and delivery speed. Also, customers will increasingly give preference to suppliers located physically near their own offices (at least in the near term before tech-nology closes the gap). Researchers have found that customers visit suppliers in inverse proportion to the distance between them. Says Russell W. Meyer, chairman of Cessna, "We spend a lot of time with subcontractors. It's a lot easier to work with someone in Wichita than with someone in Los Angeles."[22] Thus, in years to come it will not be surprising to see some suppliers actually packing up and moving closer to their biggest customer in exchange for long-term commitment. Some Japanese auto parts suppliers already have done this, setting up shop near the Nissan plant in Tennessee. The *New York Times* reports that a number of garment makers are returning to the United States from Asia for the same reason:

> Many American retailers have adopted "just-in-time" inventory controls to improve efficiencies and reduce costly inventories that have made ordering and shipping from Asia cumbersome, especially when consumer moods swing sharply or goods arrive with flaws and need to be returned for corrections. . . . The Gap, in fact, has found that it can buy American-made men's pants for about the same price as those from Hong Kong.[23]

In certain cases, small suppliers can take advantage of restrictive labor and government policies in order to succeed. The most famous example of this involves the hundreds of small factories in northern Italy. Many of these operations are family owned, having spun out of the larger customer. These firms survive by being nimble and by sparing their customers some of the tax and labor problems that are connected with doing the work in-house. As Cohen and Zysman describe it:

> Manufacturers, facing increased labor costs and restricted ability to manage flexibly inside their plants, took to subcontracting production of some components. . . . The subcontractors, being small, used less capital-intensive technology and processes than those employed within the large firms, and also, for reasons particular to Italian politics, fell outside the regulations that affected the giant companies. . . .
>
> These subcontractors often began to innovate themselves and to produce new production equipment and products. An entire sector made up of smaller firms sprang up. Public policy supported these developments by creating institutions to support these small entrepreneurial operations and to encourage vertical disintegration of production. . . . Eventually, these small producers broke loose from their subcontracting role to begin a different pattern of dynamic flexibility. They have become innovative suppliers in world markets.[24]

Companies within certain industries are forming aggregations of buyers and sellers in order to remain competitive. For example, the American semiconductor equipment manufacturers and their domestic customers have formed SEMATECH. This research consortium, backed by both the federal government and the largest U.S. independent and captive chip makers (IBM, Intel, etc.), is chartered to support the development of future generations of process equipment to keep the United States competitive. Implicit in the program is the linking of American chip makers and their domestic suppliers to a mutual destiny.

One way that both small and large suppliers have found to bind themselves more closely with buyers is to add some new value-added service to their product offerings. *Industry Week* found an example of this strategy at the General Motors Buick Reatta plant, where PPG Industries not only provides the paint for the cars but actually runs the factory paint shop. Ironically, one of PPG's tasks there is to reduce the use of its own paint. Service programs can range from repair programs (such as Caterpillar Tractor's guarantee to customers that it will deliver a part anywhere in the world within forty-eight hours), to buyer training, even to design itself.

It is important for buyers to not just passively cut supplier rolls but to actively involve themselves in "managing suppliers up to speed," to use a phrase by author David N. Burt. In many cases of apparently poor supplier performance, the real problem is actually the buyer. Thus, to drop a supplier for tardy deliveries may in fact mean both losing a good supplier and perpetuating a flawed purchasing program.

Hewlett-Packard uncovered just such a problem in its own supplier network. According to Dan Marshall, an HP manufacturing specifications manager, in 1987 only 21 percent of deliveries to the company's fifty manufacturing divisions were on time. "We wasted many hours firefighting, trying to determine which parts would be late and devising schemes to keep production lines going anyway. Early deliveries, meanwhile, were costing us a fortune in inventory storage and control." Instead of doing the usual thing—that is, blaming the suppliers—HP took a hard look at its own supplier relations. The company was amazed at what it found. A survey discovered that HP and its suppliers were clear upon their agreements only 40 percent of the time. Furthermore, HP found that suppliers had frequently shipped on time by their own criteria but had been confused by the buyer as to whether the assigned date was for shipment or delivery. Reported Marshall:

> Our study clearly revealed that *communications* was the chief culprit in on-time delivery failures—hardly a popular conclusion since it made us a primary cause of the trouble. We set out to take corrective action. . . . We changed the purchase order, labeling each date clearly.
>
> We have tried to solve the [hodgepodge routing problem] with uniform routing guides. Finally, suppliers were manually subtracting the transit time from the delivery date to calculate the ship date. We have tried to preempt such errors—and other data entry errors—by installing electronic purchase orders that flow directly from HP's computers to the supplier's open-order management systems.[25]

Within two years, 51 percent of supplier deliveries to Hewlett-Packard were on time, production was stopped less often, and inventory expenses had been cut by $9 million.

Like all types of management in the era of the virtual corporation, the management of suppliers must be built upon trust. That trust is most complete when a manufacturer allows its supplier to actually assume the task of component design. This has long been a standard procedure in the semiconductor ASIC business, but now it can even be found in pockets of America's smokestack industries. As reported in *Design News:*

> Times have changed. Today, progressive companies view their suppliers as important allies in the increasingly bitter struggle for survival and prosperity. The idea is to take full advantage of talented engineers who work for suppliers. Those skilled people effectively expand the size of your firm's design team.[26]

These words are echoed by automotive analyst Arvid Jouppi in describing why Chrysler was investing $800 million into a technology center and nearby supplier park outside of Detroit: "Chrysler doesn't have an engineer who can be dedicated to each and every component that goes into its autos. . . . By having suppliers close-by to work with Chrysler's engineers and technical people, Chrysler can continue to be an assembler of cars and use the brain trust of its suppliers."[27] The first example of Chrysler's new strategy is the Viper sportscar. Chrysler's suppliers engineered many of the car's most important components, including the transmission. Ultimately, more than 90 percent of Viper's parts came from suppliers—compared with just 70 percent for the typical Chrysler vehicle.[28]

## Balancing the Equation

The advantages to buyers of joining virtualized supplier relationships should be apparent: fewer and more skillful suppliers, higher-quality, lower-cost components arriving in a more timely manner, even a staff of design engineers to help out when needed. Add to this the less obvious advantage of being able, through greater outsourcing, to share the capital investment required to build a new product and this new type of supplier relationship looks like a terrific deal for the manufacturer.

Less apparent at first glance are the advantages to the supplier. First, it must run a gauntlet with competitors to see which will be winnowed out. It has to speed up production, consistently deliver on time, and have near-perfect production quality. On top of that, Big Brother

Buyer will be regularly making inspections, telling the supplier how to operate, tapping into its computers, and picking the brains of its most talented people. Finally, the supplier may have to drop its other customers and even change its location, all to cater to its most important buyers.

This is, of course, a one-sided view of what should be a balanced relationship. Enlightened suppliers realize that by reducing cycle times, improving deliveries, and building high-quality products they reduce their own costs and make themselves more valuable to their customers. They know that when customers invest in linking computers so they can share information, and when they become dependent on suppliers to develop products for them, they are being guaranteed a long-term source of revenue. Some suppliers, such as Baxter Healthcare (formerly American Hospital Supply), have even taken advantage of the new electronic link-ups by offering their customers added products—in Baxter's case, office supplies and even some of a competitor's products.[29]

The customer's commitment to achieving a balance can go even further. Some companies, such as BMW, sweeten the deal by rewarding their best-performing suppliers with higher margins.[30] Others pay for their suppliers' training. And some even go so far as to invest in their key suppliers—as Ford did with windshield maker Excel and diesel maker Cummins, as DEC did with chip maker MIPS, and as IBM has with more than two hundred companies throughout the world—in order to serve as a source of so-called patient capital. The result, as one Harris supplier that survived the cut explained, is that now "we get to think long-term" and "make investments that will boost quality and productivity."[31]

It is also important to understand that buyers in the new era typically will make one particularly huge sacrifice when entering a relationship with a seller: the loss of privacy. Very often suppliers will have to be made privy to company product strategies. As *Industry Week* reported: "To achieve a true partnership, customers and suppliers must share information—on new product designs, internal business plans, and long-term strategy—that once would have been closely guarded."[32]

Most difficult of all to the buyer but most important to the supplier is that the typical subcontractor relationship of the future will be single source and governed by extended contracts. Such a situation, anathema to most contemporary manufacturers, is emblematic of the new level of trust in business relationships that will be demanded by the new revolution. Co-destiny means that the buyer shows its trust in the supplier by frequently giving it exclusive production rights; conversely, the supplier puts itself in a position of extreme dependence upon the buyer for business.

Such relationships create a harmonious image of considerable appeal, certainly an improvement over the historic melodrama of the itinerant, disloyal, and price-slashing buyer and the antiquated, low-quality and unresponsive supplier. Now both, concerned for each other's welfare, at last march in step into the future. However, some of these new relationships may turn out to be an exchange of scores of small miseries for one large disaster. The supplier that hitches its wagon to a falling star, or the manufacturer that builds its business upon an incompetent sole-source supplier, will find itself part of a larger debacle, one that inflicts serious damage on both parties.

This is just one more reminder that, for all of its blessings, the new business revolution also brings with it a whole new set of dangers. The selection of partners is among the most important tasks facing the virtual corporation. When contracts are sole-source and long-term, choosing a supplier takes on all of the gravity of a joint venture—which in many ways it is. Companies moving toward such sophisticated supplier arrangements already have begun to recognize this vulnerability. For example, Xerox heavily trains its suppliers. At Ford, according to David Burt, the evaluation of potential suppliers is both detailed and, to be sure nothing is missed, multidisciplinary:

> Product development teams invite two or three qualified suppliers to compete on the design of new parts. Ford analyzes these suppliers' designs, quality plans, and price proposals. Then purchasing, with assistance from other team members, conducts a cost analysis and proceeds with negotiations. The successful proposal must satisfy a number of objectives: function, quality, aesthetics, price. The successful supplier normally becomes the only source of supply for the life of the product.[33]

Needless to say, in this scenario the supplier is also strongly motivated to help its key customer succeed. One way to do this is to make sure it is supplying that customer with exactly the product it wants. A graphic software system, called Quality Function Deployment (QFD), has been devised to help with that analysis. It tracks the material qualities the customer desires through a series of test methods to determine if the supplier's product achieves the proper parameters.[34] It is not difficult to imagine a further step, still far in the future, in which the supplier begins to look far downstream to the distribution channel and even to the final consumers themselves. A supplier might even conduct its own market research, offer its own service guarantees, even solicit design advice from end users—anything to support its customer's success.

For its part, the manufacturer must reward this supplier support with trust. Until now, firms have protected their interests by lining up multiple suppliers that could fill in for one another in case of failure. Second sources were identified in case a primary supplier failed to perform. By the same token, suppliers were proud of pointing out in annual reports how widely distributed their business was. Being dependent on a narrow customer base was considered to be a great risk.

In the business model of the future, customers will have far fewer suppliers. The just-in-case supply philosophy will be increasingly obsolete. A supplier's failure to adequately support a customer will be a serious problem. On the other hand, suppliers will be dependent on fewer customers. A customer's failure will be extremely damaging to the supplier's business.

Customers and suppliers will have few options. They will have to risk dependency in order to reap the benefits of relationships. Co-destiny is a price that will have to be paid in order to buy a ticket into the business world of the future.

## Holding On to the Customer

In a highly competitive business environment featuring abbreviated product life cycles, profits come from holding customers through several product generations. Robert Reichheld of Bain & Company estimates that retaining 2 percent more customers has the same effect on the bottom line as cutting costs by 10 percent. As a case in point, he found that for regional banks a twenty-year customer is worth 85 percent more profits than is a ten-year one. Dallas Cadillac dealer and magnate Carl Sewell has estimated that one of his lifelong customers will buy one-third of a million dollars worth of cars.[35] Thus, it becomes vital for the virtual corporation to capture customers and pull them as much as possible into a co-destiny relationship.

This cuts both ways. Consumers of all but the most basic commodities are equally motivated to establish a long-term relationship with a narrower supplier base in exchange for better service. As Robert Williams of Travelers Home Equity Service told the American Banker Conference, "People are no longer buying just products or services . . . they are buying *relationships.*"[36] To use Schonberger's words, "the customer is *in* the world class organization, not outside it."[37]

The binding of consumer to manufacturer is especially acute in the field of complex electronics products. For example, the new owner of an Apple computer quickly becomes very dependent on the manufacturer to teach him or her how to hook up the hardware and become proficient with the software. Apple can provide manuals and tutorials, but the company must also depend upon its distribution channels—such as the repairman at the local computer store—to provide such service. The trade press also helps (at arm's length, though) via magazines, newsletters, and electronic bulletin boards dedicated to Apple products. Community colleges and trade schools also do their part by offering courses in Apple computers. Needless to say, there are similar programs for IBM-compatible computers—but, underscoring the binding process, it is rare that members of the Apple "world" know much about IBM computers or vice versa. The users, in turn, by investing their time and money in learning how to operate the Apple and its software, develop a stake in the company's future.

High tech isn't the only place where consumers are developing tight, long-term relationships with suppliers. Consider the travel industry, where data bases play a key role in the kind of service the customer receives. Because it has become so simple to keep track of who the repeat customers are, it is easy to give certain ones preferential treatment. American Airlines does this with its Advantage program; as does United with its Mileage Plus. The loyal customers of these and other airlines get better seats, greater access to first-class upgrades, and more free trips. By the same token, hotels keep track of their loyal customers and make sure they get, say, a no-smoking room if they want

154

one, the right size of bed, or a room on the quiet side of the hotel.

Credit card companies are increasingly updating their product offerings also. By tracking customers' purchase and payment patterns, they are able to offer flexible credit limits to ensure that more credit is available to customers that typically have seasonal fluctuations in their credit needs.

In the automotive world, Toyota maintains up-to-date records on its customers, carefully tracking demographics to make sure the company has the right models to meet customers' needs. This means making a major investment in gathering the data and tracking the tastes of customers—and then being committed to designing products that match those preferences. Tracking the needs of its customer base is crucial to achieving Toyota's sworn objective of never losing a customer.[38] For his or her part, the consumer is also making an investment in Toyota: time spent supplying personal data and learning and analyzing the company's new products. This investment is an important factor when that consumer is ready to buy a new car. After all, if the company has a model that matches a consumer's needs, and if the trade-in value of the consumer's Toyota is higher to a Toyota dealer than it is to others, what is the likelihood of that consumer going anywhere else?

## Playing the Quality Card

One of the best tools companies can use to exceed customer expectation is quality. This has been borne out by the Profit Impact of Marketing Strategy (PIMS) studies of market strategy conducted during the last two decades by the Strategic Planning Institute. These surveys of thousands of businesses have consistently shown that quality may be the most important factor in business competition, with high product quality directly correlated to improved return on investment, profit, productivity, market share, capacity utilization, and employee morale.[39]

Quality, according to author Keki R. Bhote, is

the engine that drives a company to the bank. . . . It can be categorically stated that:

- There is no more powerful cost reduction tool than quality improvement.
- Quality is an absolute prerequisite to cycle time reduction.
- Quality is an essential step in reducing design cycle time, which is becoming one of the distinct competencies of a company in the global wars between corporations.
- Quality is one of the most important elements of—indeed a short cut to—customer satisfaction.[40]

The PIMS researchers go to great pains to note that successful quality is not that which is in conformance to design specifications but what is *perceived* as quality by the customer.[41] Ultimately then, after all the mean-time-between-failures measurements and the net-defects-per-million assembly rates, quality comes down to a subjective consideration by the customer. Which suggests that not only must the customer be involved with the company but the company must find a way to know the mind of the customer.

To be a leader in customer satisfaction will be doubly difficult— especially when competition by other virtual corporations pushes that satisfaction threshold even higher. That is why virtual corporations will require massive information-gathering systems—enlisting suppliers, distributors, wholesalers, retailers, service contractors, independent market researchers, and even the customers themselves into an insatiable quest for understanding. Says Rosabeth Moss Kanter of the *Harvard Business Review,* "The potential exists for collecting, analyzing, and using data to meet customer needs not just once but over and over again—to serve the customer without having to ask."

Such a massive (and expensive) program of information gathering will have to be, by necessity of resources and time, narrow in scope. This suggests that, as with its dealing with suppliers, the virtual corporation will have to also narrow its customer lists. It will simply be too difficult to adapt traditional strategies to selling hundreds of different product lines through scores of different channels to millions of different consumers.

Barbara Bund Jackson, in her book *Winning and Keeping Industrial Customers,* suggests one way this selection process will occur. She argues that there are two kinds of customers, defined by attitude: the lost-for-good customer, who "is either totally committed to the vendor or totally lost and committed to some other vendor," and the always-a-share customer, who "holds little vendor loyalty and can be temporarily lost or regained at any time."[42] The primary marketing task of a virtual corporation is to identify which members of its customer base belong to the first group and then do what it takes not to lose them. Nonvirtual corporations will be left to fight for the dwindling population of always-a-share customers.

## The Production of Service

Virtual products and services have a very rich service component. Much of the value of virtual products results from the ability of suppliers to make the product relatively quickly in response to demand or to custom design the product so that it precisely meets the needs of the customer. Instant delivery and custom design are both services.

Services also have a service component of their own. The core service of an airline is that of getting a customer from one location to another safely and on time. The airlines augment this with numerous other services, such as frequent-flyer upgrades, executive waiting lounges, special meals, and shorter check-in lines for certain customers.

There has been much, often sophistic, academic debate about what differentiates a product from a service. Perhaps the best definition is Texas A&M professor Leonard Berry's: "A good is an object, a device, a thing; a service is a deed, a performance, an effort. . . . It is whether the essence of what is being bought is tangible or intangible that determines its classification as a good or service."[43] Other important characteristics frequently identified include the following:

- *Services are intangible.* It is often not possible to taste, touch, or experience a service before it is purchased.
- *Services are produced and consumed almost simultaneously.* An obvious example of this is a meal in a restaurant. But it also holds true for, say, the service an attorney renders to a client in court.
- *Services cannot be inventoried.* A hospital cannot produce in advance the treatment it wishes to render to patients.

- *Customers are involved in the production of the services they receive*. A customer must be present and willing to ride on an airplane before an airline can carry that customer from one city to another.

Given these definitions, virtual products seem very servicelike. With such products, Berry's "essence of what is being bought" will often be intangible. For example, commodity products such as integrated circuits and fasteners are often identical in performance, quality, and price. What will determine whether a customer will buy from a supplier may be such things as the ability of the supplier to meet the customer's just-in-time delivery requirements. In such an instance, the essence of what is being purchased is responsiveness to customer requirements.

The goal of the supplier will be to produce products instantaneously in response to customer demand. As a customer, the virtual corporation will attempt to use the purchased products as soon as they are delivered. To the degree possible, virtual products will be consumed as they are produced, just like services. Similarly, the goal of the supplier to the virtual corporation will be to operate with as little inventory as possible. The ideal producer will have no finished-goods inventory—again, just like a service producer.

Since the virtual corporation will not produce for inventory, the output of its plant will be lost when there is no demand. The same is true for an airplane when it takes off with an empty seat or when a hotel room is not rented for a night. In all these cases the revenue-producing capability of the provider is lost and cannot be recovered.

Because virtual products are servicelike, one would expect that many of the features that characterize the relationship between service producers and their customers would be true of manufacturers and consumers as well. Two of the most important of these features are the ability of the consumer to act as a coproducer of the services he or she is to receive and the mutual competence of provider and consumer in producing the service the consumer desires. In both, an educated consumer/customer is vital.

When a virtual corporation enters into a relationship with a supplier, it takes on a great deal of responsibility to ensure the supplier's success. Perhaps the best example of this is provided by the custom-integrated circuit business, where customers are given the engineering tools that enable them to design the products they desire.

Customers who wish to receive just-in-time support from their suppliers must be intimately involved in the production of the service. That is because just-in-time support requires suppliers to produce products that are nearly perfect. (What perfection is, however, varies greatly from customer to customer. For example, one customer may be happy with components that work up to 120°F and another may require parts that work in higher temperatures.)

It is important then for the customer to understand the supplier's manufacturing process. The customer must know how quickly the supplier can respond to changes in demand. It must then provide the supplier with forecasts sufficiently stable enough so that the supplier has the chance to respond.

Equally, the consumer of a virtual product will play a far more active part in that product's creation than at any time in recent history. In the more static environment of the past, what was purchased was that which the manufacturer was willing to produce in its mass production factory. (Recall Henry Ford's remark about Model Ts being available in any color so long as it was black.) Options were rare and there was little opportunity for customer involvement in product design. Now, if the consumer is going to achieve the true benefits from the products purchased, he or she is going to have to make an investment in learning how to be an effective coproducer. This level of investment will in fact be one of the distinguishing characteristics of the relationship between suppliers and customers in the new era.

One lesson to be learned from the service industries about this process is this: it is easiest to provide good service to the most competent customer. Every doctor knows that the patient who follows directions is the easiest to treat. Every microcomputer maker knows that the most competent customers have fewer problems installing and using the equipment. Customers willing to read manuals rarely call the manufacturer for help.

Thus, one of the most important roles the supplier plays in the relationship is that of creating a competent customer. Customers have to be trained in how to use virtual products and how to interact with the virtual corporation. One would therefore expect a virtual corporation to make large investments in not only making products easy to use but in training the customers to use them.

Much of this training will happen in the normal course of business. This is another reason why stable, long-term relationships are important. They provide a great deal of time for both customers and suppliers to train one another in their needs.

There is no doubt that computers and networking will play an important role in this training process. Computer-aided instruction will make it possible for customers (and end users) to perform a considerable amount of self-instruction. Already much of the software currently sold with personal computers comes with computer programs that train the operator on their use. The multimedia systems of the future will take this process to even higher levels of interaction.

## Sharing a Destiny

For many American companies, the 1991 Christmas season served as a painful harbinger of the new business world to come. As the *Wall Street Journal* reported, many retailers, faced with the prospect of depressed sales, were "keeping tight control of inventories and forcing suppliers to swallow more risks and costs."[44] The result, the story continued, was to force suppliers to "speed up production cycles and stick to traditional products—or risk losing business."

As difficult and expensive as this was for the manufacturers, the report concluded, "some of the new pressure is good medicine. It forces them to reduce their own costs by speeding production cycles and honing their just-in-time inventory techniques. The faster response times also help American-made goods beat imports, which can't be delivered promptly without stockpiling."

Many firms, in order to survive in the middle of a deep economic recession, reluctantly began the process of creating virtual products. This placed them at an advantage over those competitors who had not made the change, but they remained behind those who had voluntarily begun the process before feeling the crush. A good example of the latter is Levi Strauss & Company, which had installed a LeviLink computer network with its major customers that enabled the jeans maker to poll retailers and replenish their inventories without even receiving an order. This quick response system enabled Levi to cut production and delivery times for a pair of jeans from forty-four days to just twenty-two. Similarly, Rubbermaid, in order to avoid what its president, Wolf Schmidt, called the "dead periods" between consumer purchases and retailer replacement orders, began tracking inventories by tapping into retailers' point-of-sale systems.[45]

Such strong business relationships will be among the most important competitive advantages the virtual corporation will possess. Unfortunately, in a confrontational, competitive, and litigious society like our own, it is hard to see how the requisite level of overall trust can be achieved.

Nevertheless, despite the odds, such relationships are slowly being constructed. This will not be an overnight process. Nor will it probably ever occur to the degree that an idealist would desire. After all, what purchasing agent, knowing a supplier's costs, will be able to resist trying to drive down that supplier's profits? And what supplier, knowing that it is a sole source, is not going to try to exploit the situation?

But the idea is to begin the process, push it forward, and let time and mutual experience slowly raise the level of trust. It will rise if only because customers, having narrowed their supply base, will have fewer suppliers to choose from and suppliers in turn will have fewer customers to sell to. The very limitation of alternatives will lead to interdependence. Over time, as the linked customers and suppliers (and linked manufacturers and end users) feel increasingly comfortable with the idea of co-destiny, relationships of trust are likely to develop. The degree to which they do will determine the ability of both parties to profit.

# 8

......................................................................

# Rethinking Management

Virtual corporation or not, the function of management is to produce results—quarterly, yearly, and long-range—that satisfy the needs of customers, employees, and shareholders and the communities in which these people live. Management that is incapable of delivering on the expectations of these widely divergent groups will not survive—and neither will the institutions for which they are responsible.

In this sense there is no difference between the management we know today and management of the virtual corporation of the future. However, the structure and methods that managers use to achieve their goals will have to change. Perhaps the most fundamental transition will be the shift that management will have to make from directing action to ensuring the smooth functioning of processes. A second change will occur in the very structure of management itself. It will become less hierarchical, and in the process much of middle management will vanish.

We know that companies that build the highest-quality products in the most efficient factories have relied on techniques such as total quality management and lean manufacturing. These techniques are in turn dependent on worker skills in problem solving and teamwork. Management's role becomes one of facilitating the processes, supporting the efforts, and taking a back seat when it comes to giving orders. The people doing the work direct their own activities. For example, former Ford CEO Donald Peterson relied on teamwork, driving responsibility down through the organization and tapping the creativity of all employees (including managers, who were more coaches and cheerleaders than autocrats) to revitalize troubled operations.[1] In the process, Ford dramatically improved the quality of its products and the productivity of its factories.

We know that the same type of management techniques are at work in companies that provide superior services to customers. Here managers have discovered that employees who have contact with customers must be empowered to respond directly to the customers' needs. Indeed, one of the keys to the service turnaround at SAS Airlines was CEO Jan Carlzon's determination in giving front-line employees the authority to meet customers' expectations without constantly having to seek approval from superiors.[2]

Much of middle management's function has been to serve as an information channel through which top managers can view events and to relay orders down to the individuals doing the work. These functions have become unnecessary because computer networks can carry much of the information about the status of operations more efficiently and effectively than can people. Also, top managers who are coaches and cheerleaders will give fewer orders because they are committed to letting the processes work and to permitting the employees to decide for themselves what should be done. If employees are making decisions about what they should do and are trained to do the right things, then there is less need to have middle management directing their activities.

It is easy to criticize these ideas as both naive and idealistic. We are so used to doing things in a certain way that the concept of a kinder and gentler management system will undoubtedly be viewed with much skepticism. But recall that today's hierarchical and directive management systems were designed to control the railroads and the mass production factories of the late nineteenth and early twentieth centuries. They are a legacy of the past, used to control a static mass production process and designed in an era when computers did not exist. These systems are as obsolete for modern flexible and responsive environments as are the buggy whip and horse-drawn carriage.

# A View of the Past

In early 1896, at the suggestion of his boss Frederick Taylor, Stanford Thompson constructed a special book for his research into piece rate assembly times. What made the book unusual was that it was actually a guise—a hole had been cut into its right-hand pages to hide a stopwatch for surreptitiously timing the activities of the workers. Great pains had been spent by Taylor in developing a stopwatch that could be operated with one hand, and Thompson showed equal care in reducing the record forms on the book's left-hand side "so that the case containing them and the watch will not attract attention."[3]

Using the stopwatch and the book, Taylor studied the various methods for performing work. His objective was to discover the one best way to do a job, upon which productivity standards could be set. Workers could be instructed in the most efficient methods, and foremen could use the standards to drive the work force to higher and higher levels of productivity. Using Taylor's scientific management techniques, it was possible to direct worker activity down to the second. The directive style of management so prevalent throughout much of current industry was born.

Maintaining secrecy was important for Taylor and his associates because they had found that workers didn't trust them, that the factory rank and file believed the American Plan (as the Taylor method was called) was just another way to squeeze more work out of them. Taylor, a former worker himself, sensed this deep mistrust. "It's a horrid life for any man," he said, "to live not being able to look any workman in the face without seeing hostility there, and a feeling that every man around you is your virtual enemy." He would add later, several years after having a nervous breakdown, "I have found that any improvement is not only opposed but aggressively and bitterly opposed by the majority of men."[4]

Of course the workers fought back, especially after labor realized that the manifold increase in production under the American Plan would not be matched with increases in salary (as Taylor had advised). In the hands of business executives who saw only greater profits by speeding up production, Taylor's scientific management, with its dream of improving the lot of both employer and employee, often became a nightmare for the latter.

John Dos Passos, describing factory life just after the First World War, wrote:

At Ford's production was improving all the time; less waste, more spotters, strawbosses, stool-pigeons (fifteen minutes for lunch, three minutes to go to the toilet, the Taylorized speedup everywhere, reach under, adjust washer, screw down bolt, shove in cotter pin, reachunder, adjustwasher, screwdown bolt, reachunderadjustscrewdownreachunderadjust) until every ounce of life was sucked into production and at night the workmen went home grey shaking husks.[5]

(In some companies, little has changed. *Rivethead,* published in 1991, tells a modern version of the same story:

As for the guy who smashed his hand, the aftermath was both sad and ridiculous. Within ten minutes of the mishap, Henry Jackson was over at the scene of the accident rantin' his fat ass off to all within earshot about how he was gonna put this individual on notice for "careless workmanship in the job place.". . . Here a man had just been permanently maimed and Henry's only concern was to see that the guy was properly penalized.)[6]

The American Plan was the first attempt to bring the tools of science and technology to the management of production. Although it resulted in a great increase in productivity, it created worker alienation. In its wake grew the schism that has set worker against management and destroyed the spirit of teamwork between labor and management that will be essential for the virtual corporation.

The second major component of today's corporate management, created originally for the railroads, is hierarchical structure. To control the railroads, management had to have information. It had to know where the locomotives and rail cars were, whether they were functional, where the freight was that needed to be picked up and where it was to be transported to, and thousands of other bits and pieces of information. This required a management-reporting structure that gathered the information, summarized it, and reported it to the next level of management, which would go through much the same process and pass the information on. Finally the data reached a decision maker who decided on a transportation strategy and gave the orders as to how to operate the system. The orders were passed back down through the layers of management and orders were given that directed worker activity.

The railroads of the late nineteenth and early twentieth centuries often functioned with towering hierarchies of management. There were managers at the center who reported to managers at the office who reported to managers at a distant location using the telegraph system. Management hierarchies made it possible to operate the railroads effectively, just as scientific management made it possible to run the factories efficiently. Combining the two created a management system that was ideal for controlling static, inflexible, mass production systems.

These management techniques were so effective that within a short time the United States was populated with the world's greatest industrial corporations. So dominant was the industrial might of our country that Jean-Jacques Servan-Schreiber, in his 1968 book *The American Challenge,* presented a vision of the industrialized world totally dominated by giant American corporations that would crush at will the industrial midgets of Europe.[7]

The great business schools of our country studied the successful industrial paradigms of the early twentieth century. The managers enshrined their successes in books and management bureaucracies. Soon thousands of managers were trained in the techniques. Generations of companies grew in the shadows of the great industrial institutions of our country, using the same methods and techniques.

A hierarchical and directive system of management might have been the ideal industrial paradigm for managing mass production in the first half of the twentieth century, but it has proven ineffective in directing the efforts of today's modern corporation. Many of the reasons for the industrial failures in the United States in steel, automotive, and consumer electronics can be directly traced to these structures. Similarly, much of the renaissance of companies such as Xerox and Ford can be directly related to the efforts of these companies to throw off the shackles of the nineteenth-century systems. The virtual corporation will have many products with short life cycles. The decision of how to respond to a rapidly changing market can best be made by the individuals closest to the action. Attempts by managers who are removed from the day-to-day turmoil of the market to direct the activities of the worker are condemned to failure.

Hierarchical and directive management will turn into a management fiasco for the virtual corporation. The system of the past, which was so effective for a static environment of mass production, will be a disaster in the fast-moving world of the virtual corporation. The principal reason for this is that levels of management mean levels of approval, and levels of approval take time. The approvers become divorced from the market. Time is the virtual corporation's most valuable resource and the one commodity it cannot afford to waste.

One cannot underestimate the role in this of the computer. Networks of computers have assumed much of the traditional role of management hierarchies. For example, railroads today mark their rail cars with machine-readable symbols and feed the information directly into computer systems, which track the location of the rolling stock. Computers can as well analyze demand and work out strategies for moving the rail cars to the right locations in order to carry the freight. Says Derek Leebaert, "The computer's pervasive effect works endlessly to force 'authority' to prove itself."[8]

If employees direct most of the tasks and computers carry the information that middle managers used to transmit, one might ask what is left for management to do? But management still has significant responsibility. It sets the goals, measures the results, directs the strategy, puts work processes in place, and establishes the environment that ensures that these processes will work effectively. For example, management in a certain virtual corporation must establish an environment in which concurrent engineering can function. It must put in place training programs, provide recognition systems, and empower employees so that total quality control can work. It must raise the level of competence so that the individuals doing the work can be trusted to do not only, in the words of management futurist Dick Cornuelle, "what they are *not told,* or even what they *can't be told.*"[9] In addition, management has to establish relationships of trust with suppliers so that just-in-time systems will function effectively.

## Going Flat

Much has been written in recent years about how corporate organization will change in the face of new communications and manufacturing technologies, global competition that demands ever-faster cycle times, and a diminishing work force. One shift is toward a flatter organization, one in which much of middle management and most of staff management will have disappeared. Peter Drucker has noted that

> from the end of World War II until the early 1980s, the trend ran toward more and more layers of management and more and more staff specialists. The trend now goes in the opposite direction.
>
> Restructuring the organization around information—something that will, of necessity, have to be done by all large businesses—invariably results in a drastic cut in the number of management levels and, with it, the number of "general" management jobs.[10]

Others see the same effect:

> New and flatter management structures become possible as more information within an organization comes on line. Organizations will no longer be forced to choose between centralization, for tighter control, and decentralization, for faster decision making. On-line technology will make it possible to have centralized control with decentralized decision making.[11]

Predicts John D. O'Brien, vice president for human resources at Borg-Warner, "I think the term 'staff function' will become extinct sometime in the 1990s."[12] Executive Dave Flansbaum is even more severe: "Middle managers are a dying breed and can, in fact, be a tremendous impediment to organizational change."[13]

Examples of this flattening of the organization chart can already be found throughout American industry. Franklin Mint, for example, has cut the number of management layers from six to four, while still doubling sales. Drucker predicts that by the mid-1990s even General Motors will have only five or six management levels, compared with the fourteen or fifteen it has now. At one time, thirteen levels of management lay between Eastman Kodak's general manager of manufacturing and the factory floor—now there are just four. Intel, already a lean company, claims to have cut management levels in some of its operating groups from ten to as little as five. And organizational consultant Jewell Westerman claims that his typical corporate client can cut the number of management layers between the CEO and front-line supervisors from twelve to six—and sometimes even to five.[14]

One of the most impressive examples of how quickly a company can be rewarded for flattening its organization is Hewlett-Packard. In 1990 the firm, historically among the premier performers in American business, was suffering from slowing sales, dwindling profits, hurt morale, and what CEO John Young called "a flawed organization mechanism." The *Wall Street Journal* diagnosed that HP "was suffering the classic symptoms of corporate gigantism: slow decision-making, sparring fiefdoms and an uncontrolled cost structure. In the fast-moving [Silicon] Valley, HP increasingly resembled a dinosaur watching fleet-footed mammals steal its nest eggs."[15]

In a tough self-appraisal, HP found that it had begun to bury itself in layers of bureaucratic red tape. Even the smallest decisions were sometimes sent all the way up the management chain to Young's office. The *Journal* reported that "to start a rebate program . . . a group manager in the laser-printer operation needed eight signatures, including those of two executive vice presidents."

Within just a year—through a hard-nosed reorganization, reduction of management approval layers, shifting of greater control to product development teams, and a more aggressive approach to the market—HP managed to turn itself around. It introduced several important new products in record time, restored its historic profit margins, and, as a result, watched its stock price climb.

One obvious result of flattening the corporate organization is that remaining managers will be forced to assume a much greater span of control. Many will have more employees reporting to them than ever before. The traditional military squad model argues that the optimum number of people that can be effectively led by one officer is about six. According to Professor J. Brian Quinn of Dartmouth, the span of control for one manager may reach two hundred.[16] This has led Drucker to argue that the corporate organization of the future will not resemble any current business models but will be structured more like "the hospital, the university, the symphony orchestra."[17]

Such expansion in span of control is simply not possible when traditional management techniques are used. There are not enough hours in a day for a manager to gather the requisite information and make informed decisions on the activities of the fifty or seventy employees author Tom Peters has suggested as the typical future span of control. Rather, new schemes of reporting and responsibility must be devised.

One of these schemes, the computer-based management information system (MIS), has been a part of most large businesses for several decades. MIS has become vital to coping with the flattening organization. Says Professor Michael Tracy of MIT's Sloan School of Management, "A sizable percentage of any organization's staff is not producing something, but coordinating something. They are in fact information conduits, types of people networks. In many cases, technology can do it better, faster. With layers of management condensed and the system dispersed throughout the corporation [MIS] offers flexibility and market-response times that preserve options. And in a time of business and general economic uncertainty, that can be a priceless edge."[18]

Technology will help with coordination and performance measurement challenges. Some companies are already implementing networks for just such a purpose. One example of this, as noted in chapter 3, is Cypress Semiconductor, which uses a companywide computer system to set ten to fifteen weekly goals for each of its fifteen hundred employees—all of which are reviewed weekly by CEO T. J. Rodgers. Crucial to this program, as even the brash Rodgers would admit, is to not let the system become oppressive and act as Big Brother looking over every worker's shoulder.[19]

Technology alone, however, is not enough to deal with the problems associated with greatly expanded spans of control and fluid organizational structures. In fact, in one crucial area—so-called thick information—management technology can be dangerously limiting. As defined by Henry Mintzberg of McGill University, thick information is irrational, subjective, intuitive knowledge that transcends what can be categorized on an MIS report, "information rich in detail and color, far beyond what can be quantified and aggregated. It must be dug out, on site, by people intimately involved with the phenomenon they wish to influence."[20]

Mintzberg's most telling explanation of such information is his comparison of the thin information used by former U.S. Defense Secretary Robert McNamara to pursue body counts in the Vietnam War with the thick knowledge obtained by U.S. ground troops merely by looking into the faces of Vietnamese peasants. Corporations (and their managers) that become too dependent upon information system surrogates for the reality of daily business life risk falling fatally out of touch. Instead, they must thicken their empirical knowledge with regular experience in the trenches with their people—yet another task on the back of the increasingly burdened manager in the virtual corporation.

Another danger involving information systems is the tendency to become overwhelmed by data—to not have the necessary software tools to sift through the mountains of figures produced by corporate information-gathering networks. Given the increasing demands on a manager's time, most will likely find themselves content to operate from summaries of summaries—in which lies another risk, that of reductionism, of summarizations that don't properly capture the true message of their source material.

To many managers, spans of control of fifty to two hundred direct reports sounds more like anarchy than Drucker's symphony. It is difficult to see how any manager working in such an environment could acquire the thick information Mintzberg suggests is necessary. It is impossible to envision a manager with one hundred direct reports being truly knowledgeable about the achievements of his subordinates and capable of writing meaningful performance appraisals. How could a manager who is spread so thinly deal with the personal problems of his subordinates or give the kind of personal recognition that is necessary?

Whether one believes spans of control can reach these extremes or not, it is certainly practical to control extremely large organizations with relatively few levels of management. For example, organizations with seven to nine direct reports at each level can reach tens of thousands of employees with five levels of management, thousands of employees with four levels, and hundreds with just three. An organization with GM's reported fifteen levels of management, along with eight direct reports at each level, could control more than thirty billion employees, or about five times the world's population.

MIS, traditionally a means of quickly getting sales, inventory, and production information to decision makers, must expand until it integrates the entire corporation. Electronic technology must be used to transfer data back and forth between sales offices, the finance department, factories, and corporate headquarters. Unless the computer can shoulder some of the work, managers will never be able to deal with the load placed upon them by the wider reporting structures. This evolution, already under way, has been reflected in nomenclature changes, as *MIS* becomes simply *IS* (information systems) to show that the passage of information is no longer unidirectional, and as portions of MIS are referred to as ESSs (executive support systems) to reflect their role in top management decision making.

171

Electronic data interchange (EDI) can do much to ease the manager's burden of interacting with suppliers, customers, and other groups within the same organizations. *Industry Week* has reported the following:

> Specifically, EDI can benefit many departments within an organization. In accounting, for instance, EDI impacts invoicing, data control, payments, electronic funds transfer, and contract progress, among others. It also enhances purchasing effectiveness through decrease in selection costs, integration with JIT inventory management, efficient order processing, and monitoring of suppliers' delivery, quality and price performance.
>
> In manufacturing, distribution, and logistics, implementing EDI can reduce inventories, foster JIT management, promote engineering data exchange, and improve work scheduling, warehouse and transport planning, and delivery notification and acknowledgement. And, in sales and marketing, improvements can be seen in product awareness, market feedback and research, reduced promotional and distribution costs, and streamlined distribution networks.[21]

Not surprisingly, IS managers have been among the first to recognize the impact of information on the corporation. Tellingly, they are among the first executives to append the word *virtual* to descriptions of their activities—as in *virtual IS,* which involves bringing crucial information instantly to the right decision maker and then transmitting the resulting decision back through the network just as quickly. Robert Morison of the Index Group writes about how corporate IS networks can find the balance between centralized, corporate-level information processing and the information needs of outlying small business units—a process he calls virtual centralization.[22]

IS makes possible wider management spans of control only if it delivers the right data to the right decision maker. It must, in the words of researcher Michael Hammer, organize around outcomes, not tasks. Historically, that has not always been the case. Says Debi Coleman, the chief information officer for Apple Computer:

> Can you imagine building the wrong product? Or finance paying the wrong set of taxes? Yet, too often IS has gotten away with following its own compass rather than contributing to the needs of the company it was created to serve. . . . Too many companies have spent millions of dollars

on the latest networking equipment and software to gather unnecessary information with an absurd degree of accuracy—only to have the CEO walk in one morning muttering, "Why can't I get today's sales data?"[23]

In the virtual corporation, the blisteringly fast cycle times and the need for instant adaptability to market changes will not allow this kind of error. Drucker is quite right when he speaks of the "information corporation," the notion that the company of the future will be organized around knowledge rather than specific products. But if that information is wrong, or gathered on the wrong subject, or sent to the wrong people, the company is worse off than if it had no IS at all.

Those who have researched the history of IS have found a pattern in the development of information systems. The implementation of a corporate information system, they have found, typically goes through three stages:

1. *The Initial Shift in Infrastructure:* A change is made in how the organization fundamentally works. Initially costs go up, but as the system gains acceptance and attains its critical mass costs begin to drop and capacity expands.

2. *The Marketing Phase:* The firms use the new infrastructure to bring out new products and services at a rate faster than their competition.

3. *The Information-Based Organization:* Information becomes a vital management support system as the application of information technology is expanded to support strategic thinking, operational decisions, exception reporting, intelligent screening procedures, and other management control processes.[24]

The most progressive American companies are racing to implement this final stage. At Apple, for example, Coleman's team is developing IS systems that reach beyond the company out to distributors to gather what she calls "channel information," data that the company needs but cannot gather by itself. She has also experimented with new kinds of presentational software, such as the so-called Slicer Dicer, to enable Apple executives to easily navigate through mountains of raw data.[25]

The computer industry is helping as well. Numerous software firms are developing IS system software that spares companies from having to devise their own from scratch. And most hardware firms are offering networking equipment for their machines.

A growing number of companies use information systems to improve quality and cut overhead and cycle times. For example, DEC saved $2 million in capital equipment costs at one pilot plant by installing an EDI system, yet still reduced inventory on select line items from $800,000 to $43,000 and cut lead times from twelve to eight weeks. HP implemented an IS system for its sales force and found an estimated 10 percent improvement in the number of purchase orders, a 35 percent increase in productivity, a cut in travel time for internal company meetings, and a jump in customer contacts. At Security Pacific Bank, the MIS system not only made possible new services (including, in a joint venture with local auto dealers, a fast financing program for car purchases) but proved so efficient that it "is now a profit center selling its services inside and outside the bank."[26]

The ultimate scenario for such an information-based organization has been described by the Index Group's John Thompson. The fully integrated firm, he says, is

> like a spreadsheet in which, when the contents of a single cell are altered, the changes automatically ripple out through the entire organization. Thus, when a customer places an order, all the related operational systems adjust accordingly: inventory, logistics, distribution plans, all the way back up the value chain into manufacturing, scheduling and beyond out to suppliers, so that the necessary parts are ordered. At the same time the systems of all the lateral functions, R&D, marketing and market research, are informed of the changes and they too "recalculate" accordingly.[27]

Even if the computer tools prove far less effective than the pundits of information technology believe, they will still enable managers to accomplish far more and deal with an ever-greater scope of responsibilities. They will provide management with a powerful arsenal of weapons with which to blast away the middle-management hierarchies that add little value to the overall management process.

# Holding Back the Tide

John Thompson's image of the adaptive corporation is an appealing one in theory, but in practice it collides with the twin obstacles of management resistance to change and unwillingness to delegate control. Effective integration of a corporationwide information system requires that management both understand the flow of data through the firm beforehand and be comfortable using technology afterward. There is considerable empirical and anecdotal evidence that many corporate managers, however, are unwilling to do either.

In 1988 an *Industry Week* survey found that while 85 percent of its readers believed that the top management of their firms would have to become more acclimated to technology in the years to come, only 54 percent thought their present management would be capable of doing so. In the same year, a private study of middle managers by Arthur Young Management Consulting Group found that executive management appeared "out of touch with reality and very unimpressed with technology."[28]

As for understanding—or trusting—the flow of company information, the same Arthur Young survey found a great divergence among the opinions of top and middle management about their companies' implementation of technology: the executives believed their company was at the cutting edge, the middle managers knew differently. One McKinsey consultant went to fifty top U.S. manufacturing sites and asked CEOs what they felt about their accounting systems:

> We asked them if, when numbers came up for an investment in automation, they believed in those numbers. Not *one* of the 50 said yes. They said, "If it gets here, somebody smart looked at it, and they massaged the numbers to be what they wanted them to be. What I make my decisions on is my gut, not the numbers."[29]

With managers unwilling to trust the information emerging from their information systems, falsely believing in the quality of their technology, and inadequately understanding how their organizations really operate—is it any wonder companies spend vast funds on IS systems that don't work? Or that they are deeply threatened by the prospect of a flattened organization and expanded spans of management? Yet, implicit in everything said thus far, there is an even greater threat to the ability of contemporary American corporations to make the transition to lean, adaptive, information-based organizations—the unwillingness of management to surrender its control.

"Knowledge is power," wrote Francis Bacon in 1597, and no one in contemporary life knows that better than the corporate manager. We have tried in this chapter to use the term *decision maker* rather than *manager* for the individual who will make the quick choices required by fast cycle times because, for the flattened organization to succeed, many, if not most, of those decision makers will have to be nonmanagerial employees.

Increased employee responsibility is crucial to the extended span of control as well, if only because it will be physically impossible for a supervisor to make all the business decisions for a large number of direct reports, all of them requiring increasingly specialized skills. Some observers predict that the resulting independence will lead to employees' coalescence into ad hoc, temporary work groups created to deal with particular problems. Says an Eastman Kodak executive, "As the span of control widens, natural teams form."[30] One of the great challenges for business in the years to come will be finding ways to make those teams work—the problem for the present is how to keep management from making those teams fail.

To allow nonmanagerial personnel to make decisions, even to temporarily form quasi-operating groups, demands that the flow of information in the company move bidirectionally—not just up the corporate hierarchy but down as well. Here we enter into the murky world of management/employee trust that frustrated Frederick Taylor a century ago. Peter C. Graham, CIM market development manager for Digital Equipment, has observed the following:

> Unfortunately, two factors still hinder the potential of networks in manufacturing. First, networking is not well understood. Our research indicates that senior management still doesn't understand the implications. The second factor is the threat that networks pose to established power bases in a company. "Where will the database reside?" "Who will have the power to make changes?" "And who will have what access to what information?"
>
> For [data networks] to be successful often depends on structural changes that permit managers to transcend their old ideas of self-interest.[31]

176

Others share Graham's view. For instance, John Leibert, president of consulting firm MDSS, believes that "the prevalence of resistance is a real and unfortunate fact. . . . Changing operations has a major emotional impact on some managers, especially when they must allow the users to enter and maintain the data that the managers had con-trolled."[32]

Ed L. Abt of Western Steel describes his experiences installing a corporate information system:

> At one place where I installed a package, there was active opposition from managers who were not about to change. They had thrived on putting out fires and were highly praised by the top management for being able to do so. They really didn't want to stop being heroes.
>
> It was almost like sabotage. They would, for instance, not mention that not all incoming materials from suppliers were inspected. The programmer, following general industry procedures, would schedule inspection for all, a needless task in this instance. Or the managers would concoct flimsy excuses not to show up for meetings.[33]

The concern of managers is not unfounded, especially in the face of statements like this from one Harvard Business School professor: "Since managers are no longer the guardians of the knowledge base, we do not need the command-control type of executive."[34] After all, many have staked their working lives on playing the corporate game by long-established rules and are now expecting the payoff—only to be told that the rules have changed and the rewards have disappeared.

"It's gut wrenching," Xerox CEO Paul Allaire told *Fortune*. "The hardest person to change is the line manager. After he's worked like a dog for five or ten years to get promoted, we have to say to him or her, 'All those reasons you wanted to be a manager? Wrong.'" What then is the new model for management in the virtual corporation? Allaire has observed that "you cannot do to your people what was done to you. You have to be a facilitator or a coach and, by the way, we're still going to hold you to the bottom line."[35]

## Learning Management

Professor Kim Clark of Harvard has written, "Executives who have thus been concerned primarily with capital investment and its return or who have put their faith in systems and procedures they thought would last forever now have to concentrate on creating a dynamic environment in which their most creative people can work hard in concert."[36] This will be even more difficult than it sounds. For one thing, the new manager's "people" in the virtual corporation will include not only traditional subordinates but also employees who have temporarily moved from other departments to be part of a task force, part-time employees, even men and women who aren't employed by the firm but work for suppliers or distributors. Setting goals for this heterogeneous collection and making sure they are met; knowing when to lead and when to stay out of the way; and merely keeping up with the group's range of interests and responsibilities will be extremely challenging.

A growing number of industry watchers are beginning to tackle the question of how to effectively manage in a virtual corporation. One popular new theory is that of participative management, the notion that employees become co-equals with their managers in planning and decision making. With participative management all the fears and hopes of managers and those they manage converge in a sort of nexus. The very notion of this type of cooperation can strike terror in the hearts of managers who relish their power and of employees who aren't interested in the burden of command. It can also, as many have noted, lead to a chaos in which every participant holds to his or her particular opinion and no one is in the position to make a final decision.

Participative management can also be perverted into a power grab by the powerless, or, conversely, into yet one more simulacrum of authoritarianism. Says Dick Cornuelle about the latter, "Many managers still think that participative management is like sandlot football where the quarterback sends *everybody* out for a pass."[37]

One individual who has closely studied the changing nature of management–employee relations is Rosabeth Moss Kanter of the *Harvard Business Review.* She has written:

> As work units become more participative and team oriented, and as professionals and knowledge workers become more prominent, the distinc-

tion between manager and non-manager begins to erode. . . . [Managers] must learn to operate without the crutch of hierarchy. Position, title, and authority are no longer adequate tools, not in a world where subordinates are encouraged to think for themselves and where managers have to work synergistically with other departments and even other companies. Success depends increasingly on tapping into sources of good ideas, on figuring out whose collaboration is needed to act on those ideas, on working with both to produce results. In short, the new managerial work implies very different ways of obtaining and using power.

Such a change, says Kanter, will force managers to find new methods for motivating their people. She has identified five such sources of motivation:

- *Mission:* Inspiring people to believe in the importance of their work.
- *Agenda Control:* Giving people the opportunity to be in control of their own careers.
- *Share of Value Creation:* Rewarding employees for their contribution to the success of the company, based upon measurable results. [This can mean skilled employees earning larger salaries than their putative superiors.]
- *Learning:* Providing people with the chance to learn new skills.
- *Reputation:* A chance to make a name for oneself in terms of public or professional recognition.

Kanter concludes by saying that "commitment to the organization still matters, but today managers build commitment by offering project opportunities. The new loyalty is not to the boss or to the company, but to projects that actualize a mission and offer challenge, growth and credit for results."[38]

## In the Eye of the Storm

What of the managers themselves? Stripped of many of the traditional perquisites of power and authority (often including even their titles), coping with an expanded and nearly impossible span of control after having lost many of their peers, working with a fluid group of employees who are, according to Kanter, "speaking up, challenging authority, and charting their own course," and perhaps even earning a higher salary, it seems reasonable to ask: what will motivate these men and women?

This is one of the most challenging questions facing the pioneers of the virtual corporation. The answer lies with the top management of each company—and most of all with the chief executive officer. The office of the chief executive of a virtual corporation is the pivot around which the entire organization will turn. Here decisions *will* be made.

One major decision will be to establish and preserve a reward system for managers that will keep them as motivated as their subordinates. This likely will resemble Kanter's proposal for employees keyed to the special needs of management professionals and harboring greater identification with the success of the corporation. Such a program necessarily involves shifting ever-greater power and independence down through the management ranks, which will most likely create a chain reaction right up to the CEO's office door, depriving him or her of much of the power a chief executive once possessed.

This will be a crucial moment for the CEO. After enforcing an apparent loss of authority and control at every step down the corporate hierarchy, can he or she make the same sacrifice? As the penultimate manager in the firm, the CEO almost by definition has the most to lose and will be the most zealous about not losing it. Can a typical CEO countenance paying one or more subordinates a comparably higher salary than he himself earns?

One unfortunate feature of American culture is that it rarely celebrates as role models the types of individuals that meet the criteria set for CEOs of the virtual company. One can think of managers and coaches of professional sports teams as rare examples of this type of management style. In military history perhaps the nearest example is Dwight Eisenhower as supreme commander of Allied forces in Europe. But neither are arresting archetypes—which, of course, is the point. Japan, by comparison, is famous for its rather invisible CEOs at the top of giant corporations.

This is not to suggest that the CEOs of virtual corporations must be anonymous figures; on the contrary, they must be very well known by their employees. Nor can they shirk the responsibilities of command. However, they can no longer solely lead the charge into battle but rather must devote themselves to developing the campaign strategy, leaving battlefield tactics to the smaller fighting units at the front.

At the same time, the CEO cannot start up the machine of the corporation and then sit back and let it run by itself down a predetermined path. The markets of the future will be too changeable for that. Furthermore, left to its own devices and with its diffused decision making, the virtual corporation risks imploding, exploding, or, most likely, careening down the wrong path. It is the job of the CEO to set the corporate vision, the corporate ethos, and to judiciously and sparingly use his or her power at the right pressure points to cause change almost invisibly.

The characteristics of such a CEO have been the focus of a number of researchers and analysts. Following are listed common themes that emerge from these writings.

- *The CEO must define the corporate vision and skillfully convey it to all employees at every level.*

"In the future," says David Luke III of Westvaco Corporation, "nothing of any consequence will be accomplished without a vision that extends the entire length and breadth of the corporation." Management consultant Fred G. Steingraber adds: "The principal responsibility for fostering [the] new organizational philosophy, including seeing the future of the company along with the customer, can only come from the CEO—from his or her values, communication, style, and way of measuring performance." And Ronald Walker of Korn/Ferry International suggests that the capacity to communicate will be a distinguishing characteristic of the executive of the twenty-first century: "Although he probably will come up through the ranks with operational experience, when he reaches the top he will have to be an effective communicator. . . . He will have to be able to tell his company's story to his employees if he wants to retain and motivate them."[39]

- *The CEO must symbolize the company.*

To customers, suppliers, and other constituents, the one constant reference point in a perpetually changing virtual corporation will be its CEO. That's one reason why Steingraber predicts that "the succession of a new CEO increasingly will be a once-in-a-generation event at most companies. . . . Lone Rangers who move from company to company, increasing their earnings and stock options at each move, will not be the pattern." A. T. Kearney researchers found that among the top-performing *Fortune* 500 firms, the average tenure for CEOs was sixteen years—while at firms that didn't perform as well the average was less than seven years.[40]

Says Ronald Walker:

> The second key quality that the 21st-century executive must possess and be able to communicate is integrity and ethics. That will be the challenge of the company's greatness. The effective executive must be able to discuss with all his constituencies what is good and bad about the company. Through his leadership, he must be his company's symbol of integrity and ethics.[41]

- *The CEO must be the company's premier generalist.*

This is the other reason why executives of virtual corporations will typically come from within the firm and will rarely jump ship. As software consultant Kenan Sahin told *Fortune,* "Before, when markets were slower, leaders had time to absorb information from experts. Now markets and technologies are becoming so complex, the experts will have to do the leading."

In truth, with decision making moved down the corporate hierarchy and the company moving rapidly to cope with changing customer needs and market conditions, it will take everything the CEO has just to keep up with change, much less propel it. And keeping up is vital, because the CEO must be sufficiently prepared to make the crucial, companywide decisions when they appear. That means the CEO must consciously place himself or herself in the frustrating position of being at the center of the corporate information system. "You've got to stay on top of the information flow," says Wilf Corrigan of LSI Logic, "while refraining most of the time from using that knowledge."

- *The CEO must trust the employees of his firm.*

Says Donald Petersen, former CEO of Ford:

> It's tough for a boss to tell his subordinates that they know more about something than he does and to run with their instincts on something. He has to have the self-confidence to trust and empower people below him in the company hierarchy. Managers who lack this confidence are reluctant to give away power, because it means they're letting go of their ability to exercise control over other people as well as what they see as proof of their personal value.[42]

With this we come full circle to the problem that has lurked at the corners of American business since the days of Frederick Winslow Taylor and decades before that. The virtual corporation is built upon unprecedented levels of trust. Between the company and its suppliers and customers. Between management and labor. Between senior and middle management.

Ultimately it comes to this: The chief executive of a virtual corporation must be able to trust employees in the firm to make responsible decisions. Those employees in turn must trust in the vision for the corporation as devised by the CEO. This is what Walker implied when he said that the top executive must be the model of integrity. And it is what John W. Gardner has meant in his writings on leadership when he says that leaders cannot maintain authority unless followers are prepared to believe in that authority, that "executives are given subordinates . . . they have to earn followers."[43]

In the virtual corporation, where almost every employee is to one degree or another a leader, the requirements for belief and trust are greater than ever before. The exact form that management in the virtual corporation will take is of course not known. We can be certain, however, that it will be different from that of most of today's corporations. The prescriptive style of scientific management—which gave minute-to-minute direction to the efforts of workers—is dead. The great management infrastructures born for the railroads in the nineteenth century are obsolete. Taking their place will be more self-management by workers, computer networks, flattened management structures, and managers focused on making processes work and establishing a vision for the company.

# 9

# A New Kind of Worker

In many ways, the General Motors plant in Oklahoma City is a perfect example of what critics find wrong with the U.S. auto industry—it builds retrograde car models using outmoded techniques. For example, most welding is still done by hand. As for automation: there are 5,300 employees and only 40 robots.

By comparison, the GM plant in Orion Township, Michigan, is technologically just what futurists would want. Its 5,600 workers build value-added luxury cars using the latest manufacturing equipment, including 170 robots.[1]

So, as the *Wall Street Journal* asked in 1991 after studying both plants, which site builds cars better?

It is the Oklahoma City plant, one product of which, the Pontiac 6000, was the highest ranking U.S. car on the J. D. Power & Associates list of the ten most trouble-free cars. By comparison, the Cadillac Fleetwoods and Oldsmobile Ninety-eights that roll out of Orion Township sometimes have as many as six defects per car—three times GM's own quality goals.

What's the difference? Why does the less sophisticated plant produce the better cars? GM Oklahoma City has implemented many Toyota production techniques, including just-in-time and lean manufacturing. But more important, its line employees exhibit a level of cooperation, identification with the quality of their work, and an esprit de corps not found at Orion Township.

For example, the *Journal* reporter was amazed to find that on any given day as many as one hundred of the Oklahoma City plant workers were in class receiving training on new equipment, computers, and basic literacy skills. What's more, these courses were taught by union people instead of management, unlike the situation at most GM facilities. At Orion Township, by comparison, the reporter found a situation of seething animosities among workers and between workers and management. In the previous few months, one employee had attacked a supervisor with a knife ("a lovers' quarrel," said management) and a worker had hit another with a power tool. Local police bitterly complained about perpetual calls to the plant to deal with similar problems.

As for the union role at Orion, the words of UAW's local chairman were guaranteed to send chills through every progressive manager: "What is quality? I mean, I have a hard time getting my arms around it." Needless to say, when it came to reading the plant's daily update on production and defect levels, the union chairman admitted to never having done so.

Beyond the fast cycle times, the lean production techniques, the implementation of new communications and data processing, after the new supplier/manufacturer/customer relationships, the adaptive organizations, and the revolutionary products, the virtual corporation comes down to the individual worker. If that man or woman has not signed on to the new business revolution, has not ratified the company's philosophy and accepted its terms of lifelong training, perpetual change, and greater responsibility, then—as the Orion example shows—no amount of new equipment or management posturing will make the slightest difference. Without the proper worker, the virtual corporation cannot even be created, much less endure.

That ratification cannot be impelled nor induced. It must be voluntary. It must occur as part of a larger redefinition of the relationship among workers, management, and, where it still exists, organized labor. This redefinition constitutes a new social contract, one that not only redistributes power but also demands an unprecedented level of cooperation and, once again, trust.[2]

Propelling this revolution in the workplace is the recognition that change has made past practices irrelevant, even destructive. As other arenas for competition—such as location—diminish in the face of the new technologies, the role of workers becomes central. "How we manage people is going to be one of the most significant competitive advantages a business can have," says Dr. Edward E. Lawler III of the USC Center for Effective Organizations, "because the traditional competitive advantages—where you manufacture and where you are headquartered—have eroded. Using people as a competitive edge is more sustainable over a longer period of time." And as researchers at Lehigh University have written, "Agile manufacturing enterprises are able to manage unpredictability by maximizing the scope for human initiative."[3]

Another reason for change is that the modern worker has begun to develop a different set of priorities relating to work. "The old motivational tools have lost their magic," according to Jude Rich of Sibson & Company. "The great challenge that lies before us is to restructure how we do work and reward people."[4]

Once again, the roots of the problem can be found in the now-obsolete rules of mass production. As reported in the *Wall Street Journal* in 1992:

> Most U.S. industrial companies still follow the precepts of "scientific management" first championed in the 1920s. . . . In that structure, managers were paid to think, and workers—often illiterate immigrants or recent migrants from rural areas—were paid to follow orders as unthinking extensions of a machine. Prof. [Thomas] Hughes [of the University of Pennsylvania] traces the current funk about U.S. work habits to managers' inability to devise a workplace suited to literate, independent-minded workers. "The values have changed; the workers have changed," he says.[5]

Just as important, the workplace revolution is occurring for conservative reasons as well. American workers enjoy some of the world's highest wages. In the face of global competition from foreign companies with far lower labor costs, worker skills must be more efficiently leveraged for their employers to remain competitive. Says Lawler, "Business must ask more of the U.S. workforce. We have to let workers contribute more and add value to the business in proportion to their wage."[6]

Lawler's view is supported by BRIE's Michael Borrus, who, based on his studies of the semiconductor industry, estimates that only 70 percent of Japan's advantage in chip making comes from better organization, while 30 percent is the result of better employee training. He says that better technology by itself won't solve the problem.[7] Michael Dell, CEO of Dell Computer Corporation, agrees: "The winners in the next few decades will be the companies with the most empowered work forces."[8]

What will daily work life be like in the virtual corporation? Tellingly, the impending changes for employees have not received the same attention as have the comparable changes for managers. Several changes do, however, appear inevitable, given the increasing technological orientation of corporations, the distribution of decision making out to the rank and file, the less distinct boundaries between the company and its suppliers and customers, and the perpetually accelerating cycle times. These changes include the following:

- More sophisticated training will continue through employees' careers.
- Cross-disciplinary organizations, such as work teams, will have extensive decision-making powers.
- Hiring policies will select for adaptability to change.
- An unprecedented emphasis will be put on retaining existing employees in a shrinking labor pool.
- Unions (where they still exist) and management will enter into different, mutually dependent relationships.
- The traditional notion of career will be redefined.
- The potential will exist for a different form of worker alienation.

## The Advancement of Learning

Throughout most of American history, the unwritten definition of *worker* included *lack of education*. The educational system was organized around this philosophy. High schools were two-tracked, with basic education and vocational training for those who would enter the working world, and college preparatory classes for those who would continue on to earn higher degrees and enter management or a profession. As written in a Hudson Institute publication, "A century ago, a high school education was thought to be superfluous for factory workers and a college degree was the mark of an academic or a lawyer. [Now] for the first time in history, a majority of all new jobs will require postsecondary education."[9]

A recognition of the failure of this model has been growing in industry for several decades, and we now find ourselves in what is generally acknowledged as a crisis in American education. At a time when U.S. corporations have an unprecedented need for a well-trained work force capable of dealing with current technologies and with the study skills needed to quickly adapt to new ones, these companies find themselves faced with hiring new employees that, after twelve or more years of schooling, are still functionally illiterate or in other ways unprepared for corporate life.

A survey by the American Business Conference of its members (primarily midsize growth companies) found that 41 percent of the firms believed that worker competence had declined—and of those respondents, 71 percent said that in response to the decline they had become more dependent upon mechanization.[10] Motorola, in the first years of its celebrated employee training program, found that it had to stop and rethink the program because it had seriously overestimated the educational level of its employees. According to William Wiggenhorn, Motorola's vice president of training, a survey of employees at a key division concluded that only 40 percent passed a test containing certain questions as simple as "Ten is what percent of 100?" Said Wiggenhorn:

> Let me dwell for a moment on the full drama of those results. The Arlington Heights work force was going to lead the company into global competition in a new technology, and 60 percent seemed to have trouble with simple arithmetic. We needed a work force capable of operating and maintaining sophisticated new equipment and facilities to a zero-defect standard, and most of them could not calculate decimals, fractions and percents.
>
> It took us several months and a number of math classes to discover that the real cause of much of this poor math performance was an inability to read or, in the case of many immigrants, to comprehend English as

a second language. Those who'd missed the simple percentage question had been unable to read the words.

Documenting installations one by one, we concluded that about half of our 25,000 manufacturing and support people in the United States failed to meet the seventh grade yardstick in English and math.[11]

The situation is expected to only get worse. Even if the educational system were to remain unchanged, demographic influences would continue to increase the general unpreparedness of the new arrivals in the work force. The baby boomers that have swelled the ranks in recent years have now, for the most part, joined the employed. This is the last group in our lifetime that will include the stereotype of the average American worker—that is, white and male. Among the twenty-five million new workers in this decade, only 15 percent will fit this profile. By comparison, 42 percent will be women, a group that historically has not migrated to technical careers; 20 percent will be native (nonimmigrant American) nonwhite men and women; and 22 percent will be immigrants, many of whom will have to learn English. The fastest growing group in the workplace will be Hispanics, an ethnic group with one of the highest high school dropout rates in American society.[12] In an aging population, none of these potential sources of labor can be ignored.

Such statistics suggest a monumental challenge. On the positive side, these groups have proven to be capable workers. For example, the same people who did so miserably on the Motorola test had managed extraordinary improvements in productivity at their plant.[13] What might they have accomplished had they not needed supervisors to translate what appeared on their computer screens? "People are much more capable than they think they are," says Ko Nishimura, president of Solectron, "and they are willing to do more than you think they will."[14]

The most progressive American corporations already assume that the average new employee will be unprepared to handle not only the special requirements of the job but probably the basic skills of reading, writing, and office interpersonal relations as well. Already, U.S. companies spend $30 billion each year on education.[15] Hewlett-Packard estimates that it spends $100 million each year on employee training and an equal amount on lost work time doing so. Similarly, Motorola invests a combined $120 million, up from just $7 million a decade ago.

Unfortunately, much of the huge training sum spent nationwide is not spent on the rank and file. *Fortune* estimates that just 12 percent of the work force receives any formal on-the-job training. So, who is being trained? Management. A survey conducted by the Rand Corporation found that while more than 60 percent of male and female professionals and nearly 50 percent of managers said they had been trained by their current employers, only a quarter of machine operators and assemblers said the same.[16] By comparison, one of the reasons for the huge international success of German *Mittelstand* companies is a sophisticated two- to four-year worker apprenticeship program. At $18,000 per year, this program may be as expensive as tuition at a top-notch U.S. university, but it also creates a highly skilled work force.[17]

## Learning to Learn

The sort of executive privilege in training found in the United States is not only contrary to the philosophy of the virtual corporation, with its empowered employees, but, what's worse, at a time when workers have the greatest need for both basic education and job skill training, it is self-destructive.

What is needed instead is continuous training for all company workers, in basic literacy and mathematics (algebra) where needed, team building, and continuing education in the new technologies appropriate for respective jobs. "The whole work force must be trained," says Sue E. Berryman, director of the National Center on Education and Employment, "and it must be continuous training, not a little splat here and there, like an injection. . . . The old idea was that the schools cooked you until you were done, and then you went to work. Now, you've got to be constantly cooking."[18]

In the words of IBM-Rochester's personnel manager Joanne McCree, "People must be 'enabled' as well as 'empowered' for this to work."[19] A number of firms have already begun to develop career-long development programs for their workers. Probably the best known of these is Motorola and its Motorola University. According to Wiggenhorn, who is president of the university, until 1980 Motorola, like most companies, "hired people to perform set tasks and didn't ask them to do a lot of thinking," and when it did train employees, "we simply taught them new techniques on top of the basic math and communication skills we supposed they brought with them from school or college."

Dwindling competitiveness and experiences like the math test at the Arlington Heights plant taught Motorola differently, says Wiggenhorn:

> Then all the rules of manufacturing and competition changed, and in our drive to change with them, we found we had to rewrite the rules of corporate training and education. We learned that line workers had to actually understand their work and their equipment, that senior management had to exemplify and reinforce new methods and skills if they were going to stick, that change had to be continuous and participative, and that education—not just instruction—was the only way to make all this occur.

As Motorola's CEO, George M. C. Fisher, succinctly put it, "We find it necessary to continually train all of our workers."[20]

Discovering just how to conduct this training was not a straightforward process. According to Wiggenhorn, the company started out with some inaccurate premises about the educational sophistication of its work force and had to backtrack quickly to make up the difference. Crucial to the survival of the program during this difficult period was the sustained support of chairman Robert Galvin, right down to personally answering complaints from employees about the training program—thus underscoring the vital role played by top management in the success of any corporate training program.

At the heart of the Motorola training program was a philosophical credo: every employee had a right to training—and retraining when technology changed. From this credo came two corollaries. One, the carrot, was that if this training failed with an individual employee because of poor pedagogy or the employee's own learning disability, it was the company's responsibility to find another way that worked. The other, the stick, was that if an employee refused the retraining when it was needed, he or she could be fired. "In fact," says Wiggenhorn, "we had refusals from 18 employees with long service, and we dismissed all but one. That sent another strong message."

In practice, Motorola has 1,200 people in some way involved with training and education. Of these, Motorola University employs 110 full-time and 300 part-time staffers. Twenty-three product design engineers serve as the equivalent of department chairs, and the senior product manager acts as a sort of dean. The university specializes in the teaching of interpersonal skills, and in recent years it has worked with local community colleges and technical schools to offer courses in basic business and technical skills such as mathematics, electronics, accounting, computer operation, and statistical process control. The program also stresses team-building skills and the inculcation of the Motorola corporate culture.[21]

Wiggenhorn likes to quote Cardinal Newman's nineteenth-century view of the ideal university—one that creates individuals who can "fill any post with credit" and "master any subject with facility"—but more pragmatically speaks of raising the average literacy of company employees from seventh to ninth grade by 1995. Says Wiggenhorn, "We not only teach people how to respond quickly to new technologies, we try to commit them to the goal of anticipating new technologies. . . . We not only teach skills, we try to breathe the very spirit of creativity and flexibility into manufacturing and management." Adds Susan Hooker of Motorola University, "We now need to keep everyone in the company going upward in terms of skill levels. The 1990s will be the decade for improving the quality of people."[22]

Motorola's is only the most celebrated of the new corporate training programs. Other corporations, recognizing changing demographics and the pressing need for a better-educated work force, have also begun to install companywide training programs. At General Electric Aircraft Engines, for example, nearly one-third of its thirty-eight thousand employees have taken a two-day course in problem solving—leading one twelve-year employee, a carpenter, to tell *Fortune,* "Before, the hourly people felt like every time we walked through the gate, we checked our brains at the guard shack. So this is starting to tap into untapped resources, which is neat."[23]

Levi Strauss, in converting one of its factories to modular manufacturing, is giving each of its workers there one hundred hours of training in such areas as reducing labor costs and laying out equipment on the shop floor. The company feels this is necessary because in the future, each employee will have as many as three different jobs.[24]

In Silicon Valley, Solectron, faced with a quarter of its two thousand employees being nonnative Americans, and many not even speaking English, created Solectron University, with courses in basic electrical assembly, American culture, and English. Said vice president Bill Yee, "The university is part of the company's cultural transformation."[25] Not far away, Intel, faced with coordinating a huge work force, spends more than $2,000 per employee each year on skill development and inculcating both workers and managers in the company's basic values regarding work ethics, risk taking, and customer orientation.[26]

One established myth is that employee training can be afforded only by large corporations because of their economies of scale. But in an age of computer terminals, VCRs, educational data bases, and thousands of community colleges, those traditional barriers no longer exist—a good thing, too, as smaller companies will most often need employee flexibility in the virtual revolution.

Amplifier maker Peavey Electronics in Mississippi found one method for training its employees: it has implemented the U.S. Army computerized Job Skills Education Program (JSEP) in conjunction with a nearby community college. Peavey embarked on the program because it recognized that its workers were not up to the demands of modern manufacturing techniques. Said Karl Haigler, special adviser to the state governor, "No one is going to take someone right off the street to run a half-million dollar robot. So this type of program is life or death for industrial expansion in Mississippi." The results of the project were staggering. Of the original sixty-four employees who took the JSEP program, a third have been promoted at least once. A second class of thirteen jumped three math grade levels in just fifty-six hours—compared with one grade level in sixty to eighty hours in traditional adult education.[27]

Plumley Companies of Tennessee, a family-owned automotive supplier, set out to train its workers in statistical process control, paying five hundred of them to attend a course at the local junior college. Like the executives of giant Motorola before him, CEO Richard Plumley was amazed to find that "over 50 percent of our workers could not add, subtract, multiply or divide. And here we were trying to teach them a first-year college course."[28]

Plumley now employs two full-time instructors and offers basic courses in reading and reading comprehension, writing, problem solving, and arithmetic. The company also offers free classroom instruction taught each month in everything from geometry and computers to three different levels of Japanese and German. In the first seven years of the program, fifty-five employees earned their high school certificates. More important for the company was the improvement in the bottom line: productivity (in sales per employee) increased 50 percent and defective products and waste dropped by more than 35 percent. Most of this, the company believes, can be credited to training.[29]

"The investment in education has more than paid for itself," concluded Plumley. "It would have been a greater cost not to make the investment. There are many people in the work force with outstanding work ethics. But many of them also need a radical change in their own self-esteem to be able to achieve their full human potential. We need to educate workers so they can respond with their heads, not their gut."[30]

One company that has set out to do just that is Lenscrafters. This half-billion dollar division of U.S. Shoe operates nearly five hundred outlets, each a factory for the real-time production of prescription eyeglasses in less than one hour. To do this, Lenscrafters has to move most daily decision making out to its geographically dispersed employees. In the process, the company has discovered a hidden obstacle to becoming a virtual corporation: the organization as a whole often does not learn from mistakes made at individual sites and thus repeats them over and over.

Says David E. Browne, Lenscrafters' thirty-two-year-old CEO:

> One of the dangers in our approach—in terms of being very decentralized and empowering—is that if we aren't capturing the key learning from the different parts of the organization, we may face a lot of redundancy. In a lot of situations, people will face a problem for the first time. As a company, we may have faced it 20 times and figured out four different ways to attack it. . . . There's a major efficiency to be had by having everybody understand what we've learned to date.[31]

The Lenscrafters dilemma is a reminder that the process of becoming a virtual corporation, before and after everything, is about learning. The virtual product itself is a learning machine, gathering information to be used in the design of its successor. By the same token, the virtual corporation is a learning entity, struggling to understand its mercurial operating environment so as to successfully adapt to it.

The individuals who make up the virtual corporation—employees, contractors, even suppliers and customers—more than anything else must be full-time learners.[32] This doesn't mean "trained." Simple skill development is not enough for the continuous and radical changes of virtualized business. Any such skill could quickly be rendered obsolete or irrelevant. Rather, participants must learn how to learn. They must be equipped with the conceptual skills required to deal with perpetual change. And they must be armed with the technology needed to put this ability to work.

One way Lenscrafters helps this learning process along is by making the acceptance of mistakes one of the company's core values. Says Browne, "It's OK to fail in our [corporate] culture as long as you try ideas and have something not work, as long as you learn from it and the company learns from it. . . . Accepting mistakes is important. It removes fear. It encourages innovation. There's a lot of room for folks to tailor-market, to tailor merchandise, to tailor their operations approach."[33]

Progressive European companies are also pursuing employee education. BMW, for one, regularly sends its forepersons to Study City to learn total quality production skills.[34] A nontraditional European company that has tapped into this learner market is The Body Shop Corporation, a British natural cosmetics firm with franchises throughout the world and with annual revenues predicted to reach $1 billion by 1995. Despite its countercultural airs and founder Anita Roddick's sworn hatred of Harvard MBAs, The Body Shop has long recognized the importance of training and information as employee motivators. The company's training center in London is open to any employee, not only of the parent firm but of every franchiser around the world. As Roddick told *Inc.*, other cosmetics companies "train for a sale. We train for knowledge."[35]

## The New Breed

One striking feature of the manufacturing and office floors of the virtual corporation will be how little the employees will actually look like their counterparts of our time. At the turn of the century, 45 percent of the U.S. work force will be white males, down from 47 percent in 1990. But the types of white males will be quite different.[36] The most eligible of the baby boomers have already been picked, as will all in the subsequent generation as soon as they leave school.

That leaves those white males who for some reason or another—low intelligence, physical disabilities, criminal record, emotional dysfunctionality—have opted out or been excluded from the workplace. It also leaves the perpetual have-nots of American work life, black males for example, who for a number of reasons—high dropout rate, lack of workplace experience, racism—actually risk seeing their presence in the work force recede. In the labor crunch that will come from an aging population, companies will be forced to take a second look at these groups and find ways to accommodate them in the workplace.

One group that will show an increasing presence in the U.S. work force will be women. Native (nonimmigrant American) white women will represent the fastest growing fraction of the labor force, contributing an estimated 42 percent of the new workers between 1985 and 2000. Native nonwhite women, especially black women, will also hold a larger fraction of jobs in the years to come. These figures, combined with those of immigrant women, suggest that the percentage of females in the labor force will, for the first time, approach that of men in the labor force, and these women will contribute almost two-thirds of the new workers in the near future.[37] This constitutes a powerful interest group that may force many hungry companies to adopt as recruiting tools so-called women's policies, such as extended maternity leave, sick child leave, part-time and home work, and on-site day care.

Immigrants will remain one of the most dynamic components of the labor force, with more immigrant men (13 percent) and immigrant women (9 percent) added to the work force before the end of the century than native nonwhite men (7 percent). Immigrant workers present their own unique problems, most notably in the inability to speak English, a presence of cultural confusion, and, for many refugee-type immigrants, a lack of training in the tools of everyday modern life.[38]

In the past, corporations might have chosen to ignore these groups, arguing that their job was to make a profit for investors, not subsidize social change. That strategy is increasingly untenable. In accepting the challenge of hiring from less-adapted groups, companies must also accept that many established mores of the office and factory must go by the wayside. Barbara Shimko of Widener University offers a glimpse of the magnitude of this task:

It is understandable that recruiters want to hire persons from identifiable mainstream groups who in the past have proven to be successful, easy-to-manage employees. The applicants that recruiters are attracting these days do not belong to those desired groups. These include, among others: minorities, females, older people, handicapped, disadvantaged, and those lacking work experience. In addition, some applicants will have the added burden of a major personal problem such as a discordant home life (if they have a home), a prison record, or a history of alcohol or drug abuse. In all likelihood, if you hire someone and then decide to fire him or her because things are not working out the way you had hoped they would, the person you bring in as a replacement is going to seem very much like the one you just let go.[39]

Unfortunately, of all the hiring criteria used by American corporations, many are more social tics than anything else, designed—consciously or not—precisely to keep out these potential sources of labor. For example, Shimko and her group surveyed thirty-eight general managers of fast-food restaurants to determine what they felt were crucial factors in the hiring or not hiring of job applicants. Beyond the usual matters of attitude, prior experience, and honesty, Shimko found that 20 percent of the managers considered appearance a factor in the decision not to hire (tied for the highest percentage with those who considered "attitude" a factor). In particular, this meant hairstyles, jewelry, and brightly colored or unusual clothing the manager found objectionable—characteristics that might simply signal a different cultural upbringing.[40]

But most amazing to Shimko was that 9 percent of those receiving a negative hiring decision were turned down for inappropriate eye contact:

> To give a firm handshake and look someone straight in the eyes is a very important lesson taught by Dad to every middle class male at a tender age. Not only do non-mainstream groups miss the lesson from Dad, some are taught that direct eye contact is rude or worse. Girls are frequently taught that direct eye contact is unbecoming in a female. In reality, having averted or shifty eyes may indicate mostly that the job applicant is not a middle class male.[41]

Psychological researchers in the mid-1970s found that many of these new types of hires had not had a proper job as an adult and thus misread as threats what were in fact standard business practice. For example, to some workers, a dressing down by a supervisor might be seen as abuse or prejudice, thus some employees may respond with deep distrust and resentment and fight back with petty theft, surliness, or idleness. This employee is then labeled a "behavior problem" and the cycle spirals down to angry termination or even violence.[42] Needless to say, this behavior isn't unique to nontraditional hires. And, as with appearance and office behavior, a little preemptive training can preclude workplace tension or even tragedy.

All of these caveats aside, the new, nontraditional workers often turn out to be important contributors. For example, nontraditional employee groups often show dedication, loyalty, and even special skills not typically found in traditional employee groups. Marriott Corporation, suffering from a 105 percent annual turnover rate among its workers, initiated a program to hire physically and mentally disabled workers. Though the program had its costs—special social and job training courses, as well as teaming the new hires with company managers—the turnover rate of this group is now just 8 percent.[43]

In Philadelphia, Project Transition trains welfare recipients specifically for the fast-food industry, as it is not only one of the fastest growing U.S. industries but also has one of the worst turnover rates—and thus would be more receptive to hiring Project Transition graduates. Given the dangers of manager prejudice, one of the things the project teaches its participants is how to impersonate mainstream job applicants during interviews. To help the newly hired deal with their unpreparedness for work culture, project graduates are also helped with eight weeks of on-the-job coaching.[44]

# Teamwork

The volatile nature of the virtual corporation will be reflected throughout its organization. If this is true in the ranks of management, with its shifting spans of control, it will be even more so among workers.

The empowerment of employees, combined with the cross-disciplinary nature of virtual products, will demand a perpetual mixing and matching of individuals with unique skills. These individuals, as their talents fit, will coalesce around a particular task, and when that task is completed will again separate to reform in a new configuration around

the next task. The effect will be something like atoms temporarily joining together to form molecules, then breaking up to form a whole new
set of bonds.

In the virtual corporation it will not be unlikely for a task force to
form, almost spontaneously, around a common project. Such a group
might contain representatives from the research lab, the manufacturing floor, a sales office, even from suppliers, distributors, academia—
and the customer. This group might meet on a regular basis, but often
would be geographically dispersed and would communicate using tele-
phones, computers, and electronic mail.

In recent years, the most discussed example of this sort of group
activity is that of the work team. In *Fortune*'s description, work teams

> typically consist of between three and 30 workers—sometimes blue collar,
> sometimes white collar, sometimes both. In a few cases, they have become a
> permanent part of the work force. In others, management assembles the
> team for a few months or years to develop a new product or solve a particular
> problem. Companies that use them—and they work as well in service or
> finance businesses as they do in manufacturing—usually see productivity rise
> dramatically. That's because teams composed of people with different skills,
> from different parts of the company, can swoop around bureaucratic obsta-
> cles and break through walls separating different functions to get a job done.[45]

The magazine went on to note that a survey by the American Productivity and Quality Center found that half the 476 *Fortune* 1000
companies it had surveyed planned to use work teams in the future—
but that, as of 1990, only half had done so. Among those that did organize around work teams, the improvements have been impressive:

- So successful have teamwork programs been at the Defense
  Systems and Electronics Group at Texas Instruments that management announced the goal of having every employee in a self-
  directed work team by the end of 1992.[46]
- The General Mills cereal plant in Lodi, California, runs during
  the night shift with no managers present; work teams at the
  plant have increased productivity 40 percent.
- A Federal Express work team identified a billing problem that
  was costing the company $2.1 million per year.

- Aetna Life & Casualty reduced the ratio of middle managers at its home office from 1:7 to 1:20 while still improving customer service.
- Work teams at Johnsonville Foods in Wisconsin convinced their CEO to make a major plant expansion, and the result has been a 50 percent improvement in productivity.[47]
- Corning Glass took the bold step of organizing 70 percent of its twelve hundred scientists and engineers into quality work teams, an unheard-of notion in R&D. The results in the first four years included savings of more than $21 million, faster new product creation, and a doubling of Corning's return on equity.[48]

There are several obvious advantages to work teams. For one, they reduce the need for layer upon layer of middle management. They also move control down to the people with the most obvious, hands-on experience with the process.

In the factory, some managers believe that with work teams the costly creation of individual labor standards can be dropped in lieu of estimates of required labor for the total operation. It also increases efficiency. For example, operators don't have to stay at their work areas if they finish a task—they can go help others with their jobs. The process is self-motivating and self-disciplining, as the team begins to develop an esprit de corps and pulls along its laggards.[49]

It is important to note that work teams in themselves are not a solution. In fact, one survey of the auto industry found that teams can actually reduce productivity. The real gain, these researchers found, occurred when team organization was linked with greater participation in shop floor decision making.[50] Another survey found that team members listed the following as the three most important benefits to working in teams: improved involvement and performance, positive morale, and "the sense of ownership and commitment to the product that teams create."[51]

Work teams are a prelude to the kind of ephemeral organizational structures that will be found in the virtual corporation. The aforementioned survey found that the typical team size is six to ten members and that three-quarters have been in existence less than three years; a little more than half have been around for less than two years.

From a team of five assembly line workers in a factory to a product design team of fifty scientists, designers, production experts, and customers spread throughout the world in a dozen different time zones is not an easy jump. The great challenge to teams in the virtual corporation will be overcoming the limitations of both geography and time. Team members may not always be able to communicate directly, much less in person, so systems must be in place that support collaboration through other means, such as computer data bases. Apple Computer is currently working on just such a program, a so-called electronic campus that would allow team members to interact through common data bases that organize contributions not sequentially but thematically.

Developing productive teams, even of the basic variety, is not simple. First, not every job merits a team approach. Ordinary, solitary tasks usually don't benefit from this type of reorganization. Second, as writer Michael Schrage explains in *Shared Minds*, bringing people together in a team structure does not guarantee fruitful collaboration:

> When it comes to human communication, there's a factor more influential in everyday life than most people care to admit. . . . It's our ability to deceive ourselves. . . . This self-deception runs through most organizational communications. It's why people think meetings are a waste of time, and it helps explain the frustration most people feel when they try to collaborate.
>
> Most people kid themselves into thinking that they're collaborating with someone when, in reality, they're just saying words. Traditional modes of discourse in no way capture the subtleties, the power, and the degrees of interaction necessary for effective communication.[52]

Students of work teams have discovered them to be far more complex in their behavior than one might expect. Researcher Glenn Parker found that a successful team typically exhibits a dozen different traits, from a clear purpose to ruthless self-appraisal. He also identified four types of team player: the contributor, who is information and performance focused; the collaborator, who is the source of the team's vision; the communicator, who is the facilitating heart of the team; and the challenger, who serves as the team's devil's advocate. Every team member exhibits one or more of these traits, and a great team includes members who make sure that all four roles are being performed in the right proportions.[53]

Allan Cox takes this one step further, defining an effective work team as

> a thinking organism where problems are named, assumptions challenged, alternatives generated, consequences assessed, priorities set, admissions made, competitors evaluated, missions validated, goals tested, hopes ventured, fears anticipated, successes expected, vulnerabilities expressed, contributions praised, absurdities tolerated, withdrawals noticed, victories celebrated, and defeats overcome.[54]

Obviously, this is a much more sophisticated vision of a work group than a glorified quality circle or a task force to determine employee vacation schedules. Add temporal and geographic elements to Cox's definition—that such a team may be ephemeral or nearly permanent, that it may have a handful of members or dozens, that its participants might not all be employees, and that they may be dispersed throughout the globe—and we have an image of work organization in the virtual corporation.

Reaching this more sophisticated level of teamwork will require not only more advanced hardware and software, but also much more training of the individual team member.

## Learning Together

If this chapter has a single dominant message, it is that in the modern business world employee training becomes paramount. This means teaching basic education and social skills in which employees are lacking. Because the corporation of the future will be built on information, it means that it will be necessary to educate people about the tools that control and manipulate information. And, because teamwork will be the primary work mode, it means that training in consensus building, group dynamics, and problem solving will be essential.

This is not an inexpensive process. Unfortunately, most companies simply march into a work team organization without the requisite employee preparation. One company that learned its lesson early was Corning. In 1983, when it set out to improve quality through the use of teamwork, its management established two areas of training emphasis: statistical process control and problem solving. A year later, after a fitful start, the company realized it also had to add a course called Group Dynamics and Communication.[55]

Working as a team is natural human behavior, but dealing with the nuances of targeted teamwork involving several diverse personalities and skills is a tricky business, one for which most employees (and many managers) are unprepared. "Upper management has to be prepared to spend some bucks," says Russ Preston, first-line supervisor at Fisher Controls. "It has to put forth the money if it wants work teams to succeed."[56]

Unfortunately, although studies show that a near majority of business executives predict that half or more of their work force will be organized as self-directed teams, willingness to pay the cost in money and loss of control seems to be lacking. For example, an estimated one-third of teams aren't even allowed to select their team leaders. In 55 percent of the teams, management retains all rights to prepare and manage the team's budget—and only 7 percent let teams make their own compensation decisions, while 59 percent still cling to individual merit pay programs.[57]

Managers themselves recognize the problem. When asked, 54 percent admitted that the greatest barrier to successful work teams was insufficient training, followed by supervisory resistance, incompatible computer systems, lack of planning, and lack of management support.[58] It is this kind of result that underscores Peter Drucker's well-known observation that management spends much of its time getting in the way.

The virtual corporation is a learning organization. At any given moment it is a collection of skills, talents, and experiences that reside in the minds of its managers and workers, and a body of information relating to its products, its internal structure, and its business relationships. Those skills, talents, and experiences bear upon that information—analyzing it, packaging it, and using it to improve the firm. To do this requires basic skill levels and career-long training and retraining of all employees. For a company to survive in the new business environment, all of its employees must learn together. Wrote the Lehigh researchers, "The quality of all manufacturing jobs, including those in production, will be enhanced by the premium placed on initiative, knowledge, and active involvement in all levels of the manufacturing organization in setting and executing production agendas."[59]

Earlier we spoke of a new social contract, one built on a spirit of cooperation and trust. Training is part of management's contribution to this contract, along with greater employee power and a just reward system that reflects this added responsibility. Labor's contribution is increased productivity, a willingness to contribute the time needed for added training, and the assumption of much of the role of the old layers of middle management.

An example of how cooperation and trust can yield remarkable results can be found at an experimental plant opened by Cincinnati Milacron in 1991. There, work teams built complex, computer-controlled lathes without supervision or even time cards. Absenteeism is almost zero and worker complaints have fallen sharply because, as one fourteen-year veteran told the *Wall Street Journal,* "It makes you feel like you're really part of what you're doing, instead of management telling you to put this bolt here and that bolt there. [Now], if you need to stay a half-hour longer, it's never any problem."[60]

There are secondary implications of this new social contract as well. One of the most important is the rise of a different sort of interdependence between worker and employer. The employee becomes increasingly indispensable. Not only will a considerable sum have been spent upon that person's training but, thanks to the added responsibility, the employee (or team) will know more about how to, say, sell company A's products to customer B than anyone else. The amount of time and money required to replace such an employee will be large. Thus, de facto, guaranteed, long-term employment will be a common characteristic of the virtual corporation. Some companies have already discovered this change. At LSI Logic, in the Silicon Valley (a region with a history of a migratory professional work force), CEO Wilf Corrigan admits having adopted a policy of doing everything reasonable to retain employees. "Trained people are just too hard to replace," he says.[61]

But this employee leverage is matched by a countering force. Many of these new skills will be so specialized as to be nontransferable. The same skill at selling A's products to customer B that makes the employee all but irreplaceable to company A also makes that employee of less interest to company C, its competitor—and even less so if C doesn't sell products to B. Thus, a new balance of power evolves, one that forces the employer to shift ever-greater authority to the employee and in return makes that employee vitally involved in the long-term survival of the firm. Conversely, the employee loses a lot of his or her flexibility in the job market.

Combine worker irreplaceability with changing demographics and it becomes apparent why many forward-thinking firms are already adopting new kinds of benefit packages and employment practices such as maternity leave, on-site child care, cultural support groups, and job sharing in order to keep the emerging nontraditional work force happy and loyal.

## Armed Camps

One might well ask where organized labor belongs in this new scheme. Certainly, cooperation and trust, the watchwords of the virtual corporation, are hardly terms that would be used to characterize the history of labor/management relations—at least not since the Civil War and the rise of the railroad, postal, and telegraph industries.

These enterprises, as they grew in scope and size, began to devise the power alignments that still define the modern corporation. The most notable of these was that of middle management and its assumption of duties that previously had taken place on the factory floor. Says Alfred Chandler:

> With the coming of the modern factory, the plant manager and his staff took over from the foreman the decisions concerning hiring, firing, and promotion, as well as those on wages, hours and conditions of work. As the enterprise grew, such decisions were placed in the hands of middle management. Policy matters were determined by executives in new personnel departments housed in the central office.[62]

Part of this change was due to necessity. By the end of the nineteenth century, many corporations were growing simply too large to be managed by locally improvised rules. In 1891 the Pennsylvania Railroad, for example, had more than 110,000 workers—twice the number in the U.S. armed forces at the time—and had revenues almost a third as great as the federal government.[63]

A second reason for the evolution of middle management was philosophical. The myth of the capitalist plutocrat aside, even by the turn of the century most large U.S. enterprises were not family or financier owned but were public corporations with a professional management.

This management considered efficiency and margins rather than empire as the prime forces in the adoption of Taylor's American Plan and other so-called scientific methods of production and organization.

Whatever the improvements in coordination and profitability, one inevitable by-product of the bureaucratic corporation was worker alienation. A schism between labor and management had always been intrinsic to the nature of business hierarchies, but in the modern corporation that split became institutionalized.

The worker was the net loser in this power shift. Having already lost most of the control over the product of his labors in the transition from craftsmanship to mass production a few decades earlier, the worker now saw the erosion of his control over the organizational issues of daily work life.

The opening decades of the twentieth century saw the struggle of large segments of the labor force to retain, or regain, power in the workplace. The most common means was through unionization. Despite the legends of early labor battles, most of the organizing successes occurred in industries where modern corporate life had yet to take hold. As Chandler tells it:

> Except on the railroads, the influence of the working force on the decisions made by managers of modern business enterprises did not begin until the 1930s. Before then craft unions had some success in organizing the workers in such labor-intensive skilled trades as cigar, garment, hat, and stove making, shipbuilding, and coal mining—trades in which modern business enterprise rarely flourished. . . . The craft unions, however, made little effort to unionize those industries where administrative coordination paid off. . . . Until the 1930s, [middle managers] were rarely forced to consider seriously the demands of labor unions to represent the workers.[64]

The change did come in the 1930s when, through groups such as the Congress of Industrial Organizations (CIO), unions began to organize along industrial rather than geographic lines. Now labor could answer the specific needs of unskilled or semiskilled workers according to the nature of the work itself and use the leverage of cross-country, industrywide strikes. For the next three decades, U.S. industrial unions made impressive gains.

But these near-term successes came with serious long-term costs,

costs that are now being paid by unions experiencing dwindling rolls and influence. In particular, unions and management codified the nature of their relationship: mutual distrust, an ongoing strategy of most companies to drive out their unions, a reactionary resistance to technological and structural change by the rank and file, and most deadly, a willful ignorance by union leaders about the larger issues facing the corporations where their members worked. "Management proposes and union disposes," as the saying went, and in the process most American trade unions yielded control over the fate of the firm to corporate management. Chandler wrote:

> Union leaders, during the great organizing drives of the late 1930s and immediately after World War II, rarely, if ever sought to have a say in the determination of policies other than those that directly affected the lives of their members. They wanted to take part only in those concerning wages, hours, working rules, hiring, firing and promotion. Even the unsuccessful demand "to look at the company's books" was viewed as a way to assure union members that they were receiving a fair share of the income generated by the company. The union members almost never asked to participate in decisions concerning output, pricing, scheduling, and resource allocation.[65]

By the 1980s, union membership in the United States was experiencing a precipitous drop. From a high of more than 35 percent of the work force in 1955, it had fallen by 1989 to just 12 percent (and still only 16 percent if one included government employees), the lowest rate since the early 1930s.[66]

There is little indication that this situation is going to improve. One problem is that unions are suffering from their past successes. Calling unions "this century's most successful institution," Drucker goes on to suggest that unions have reached a point of diminishing returns:

> The labor union certainly has much less to offer. Practically everything it stood for has become law in developed countries: short working hours, overtime pay, paid vacations, retirement pensions, and so on. The wage fund, that is, that part of the gross national product that goes to employee, now exceeds 80 or 85 percent in all developed countries. This means that there is no more "more" for union members. In most years the

employer contribution to the employee's pension fund already exceeds by a good margin all the profit available to the shareholder.[67]

Management, forever fearful of union power, has happily helped in its demise. "Management autocracy is on the rise," says Harry Katz of Cornell. Labor negotiator Harold Hoffman adds, "Many managements don't want the unions to be part of the corporation's moving towards the future."[68]

These and other factors—the declining quality of union leadership, an unsupportive Republican administration in Washington, and the shift of manufacturing to the traditionally nonunion South and West—are all partial explanations for the decline of unions in the United States. But the precipitous decline, one not shared by other industrialized nations (union membership in the United Kingdom and Japan has dropped only slightly, in Germany and Canada has remained even, and in Sweden has jumped from 79 percent to 96 percent), suggests that the primary problem with American unions is internal. It is this internal misorientation that has kept unions from making gains at a time historically considered most fertile for worker organization—an era of declining standards of living, record unfair labor practice complaints, and widespread worker dislocation.

The most common explanation for this decline is that labor in the United States has lost touch with the changing realities of the workplace—a dangerous position to be in as we approach the radical discontinuity of a new business revolution. Wrote John Hoerr in the *Harvard Business Review:*

> Ask a manager to explain this seeming paradox between worker discontent and union decline and it's likely he or she will say something like the following: the rules of the economic game have changed. Competition is global, technological innovation continuous, the workforce increasingly professional. In such an economic environment, unions are ill-suited to meeting the needs of either workers or companies. At best, they are an irrelevance—a leftover from a previous industrial era. At worst, they are an obstacle to making companies and countries competitive. Little wonder, then, that unions are on the wane.[69]

A survey of pro-labor articles, books, and speeches is equally dispiriting, leaving one with the impression that organized labor in the

United States is still fighting the battles of the thirties, with little recognition—if not paranoia—about the changes taking place within its membership. Labor writer Jane Slaughter (*Choosing Sides: Unions and the Team Concept*) argues that work teams are merely the latest management subterfuge, a new form of worker speed-up, and decries the willingness of some unions to work with the "employer class" in participatory management schemes as a form of co-option.[70] Radical education writer Jonathan Kozol sees business participation in schools in low-income areas as a cynical attempt to increase profitability by filling future low-paying jobs.[71]

American organized labor has backed itself into a corner by removing itself from corporate decision making and instead focusing upon enforcing—through job classifications, seniority schemes, and so on—organizational rigidity. This is the most suicidal strategy imaginable in the virtual corporation. That is why U.S. unions, in the words of Hoerr, "must reinvent themselves much as some companies are trying to do."[72]

Corporate managers who might welcome the end of unions are making a potentially dangerous mistake. Labor will be represented in one form or another. The void created by the disappearance of unions would probably be filled by government statutes and regulations of far less flexibility than the average union negotiator. And executives should also note that, as the pro-union Economic Policy Institute did in early 1992, "if unionization were a necessary condition for our declining competitiveness, then non-union industries should be unaffected. The declining competitive position of non-union high-tech industries in the late 1980s makes it clear that there is no necessary connection."[73]

There is a positive reason for bolstering labor unions in some industries as well: if sufficiently enlightened, they would actually enhance the higher levels of training and teamwork required in the virtual corporation. At places such as Corning Glass, union participation in work team reorganization and training has proven to be a powerful tool. A Carnegie-Mellon/MIT study of one thousand manufacturing plants found union shops with work teams were actually more efficient than their nonunion counterparts. Wrote the study's authors: "Ironically, it is precisely because unionized workers can say 'no' as a group that they can also collectively say 'yes.'"[74]

One place American labor can look for new models for organizing labor is in other industrialized countries. After all, in Europe and

Japan, unions remain a vital presence in the society—and though some traditional measures of union power have been lost in those countries, others have been gained.

Lowell Turner of Cornell has studied unions around the world. His conclusion is that the varying successes of these organizations are dependent upon how well the representation of worker interest has been institutionalized. "In particular," he says, "two critical variables account for relative union success or decline and the stability of industrial relations systems in the contemporary period: (1) the extent to which unions, as a broad national pattern, are integrated into processes of managerial decision making; and (2) the relative cohesiveness of the national labor movement."[75]

How unions have dealt with these challenges has been different in each country. In the most extreme case, Japanese labor is organized into enterprise unions, that is, company unions in which the leadership not only works closely with the company but often becomes an integral part of management after a certain tenure. Needless to say, such a system would be anathema to adversarial American unions.

Far more transferable, Turner found, was the West German model, which operates through a system of legally empowered works councils that are independent of both union and management and are elected by the work force, yet have strong union connections. These works councils are participants in personnel, training, and reorganization decisions made by their companies and by law are privy to the introduction of new technology and job design.

This subtle interplay between works council, union, and management can slow decision making but it builds consensus. How it operates is exemplified by the German Metalworkers Union (IG Metal). IG Metal, recognizing the external challenge to the German auto industry by Japanese competition, took an aggressive tack, using its membership strength to wring concessions out of management in exchange for supporting the new work team organization. The result was that the union "held on to its high membership levels and density, expanded protections for its members, retained a stable position of influence within the plants through union-dominated works councils, and began to promote its own vision of work reorganization as part of a general pattern of plant-level productivity coalitions throughout the auto industry."[76]

It is important to recognize that IG Metal actually led the push in its industry for work teams, having preceded management in developing both a vision for this new type of organization and a strategy for its implementation.

The impact of this approach to union management relations can be found in the success of Volkswagen in the world market. Writes Turner:

> The key elements of the [VW] model are: cooperative or "social partnership" relations between labor and management; the virtual identity of unions and works council; considerable engagement of the works council in managerial decision-making processes; unity within the works council and union, so that differences regarding such critical issues as policy and candidate selection are hammered out internally and a united front is presented in negotiations with management; a high rate of union membership and strong union shop floor presence (over 1,000 shop stewards at Wolfsburg alone); virtual lifetime pay and employment security for the workforce; a management (from top to bottom) that is trained to listen to the concerns of workforce representation and to seek consensus prior to the implementation of policy; and last but not least, a firm that is highly successful in world markets, whose management and labor representatives at least in the past have regarded "cooperative conflict resolution" at VW as a source of competitive advantage in the marketplace.[77]

Elsewhere in the world, Turner found unions, all faced with a changing competitive environment, either adapting or fading. In Italy, labor and management had long operated as adversaries, each taking advantage of shifts in the balance of power. Fiat, as the paradigmatic case, had spent much of the 1970s accommodating to union demands in the face of strikes. Then, in 1980, as the company faced potential bankruptcy, Fiat's management went on the offensive. By the end of the decade, it had broken the union, reducing it to a marginal role as it set about implementing new organization and worker participation schemes.[78]

Richard Locke of MIT has also studied unions in Italy. In his opinion, the national unions have failed because they cannot deal with the many diverse reorganizations being attempted by Italian companies. The reaction of the national unions has been one of brute force, that of trying to impose a monolithic national labor policy. "These vertical structures appear unable to adapt to the variety of corporate structures and strategies emerging within their sectors," writes Locke.[79]

Beyond that, however, Locke has detected the rise of local unions, notably in the northern Biellese textile district. Despite enormous difficulties—for example, its members are communist and company management is rightist—these small, focused unions have worked with many local companies to help them reorganize and prepare for the transition to specialized production and to implement new technologies. Says Locke:

> As one local business leader put it, the unions and the managers united in a "pact for development" in order to save the local industry and preserve jobs. . . . Cooperation continues between unions and business leaders. Joint efforts have emerged to promote research and development, technical education and job retraining, and improved infrastructures—all aimed at enhancing the competitiveness of local industry. The results have been positive. Record sales and profits rates for firms have been matched by high rates of employment.[80]

In Sweden, the most organized of all Western labor forces faced its own challenge in the 1970s. Having historically maintained an arm's-length relationship with management, it presciently recognized the new forms of organization as both inevitable and a threat to marginalize its own future role in society. Having strong government support, it turned to legislation and new collective bargaining agreements that emphasized "co-determination" of company strategy between management and the union. This not only precluded future management challenges but gave the union a new role in defining workplace organization and employee training. As a result, as noted earlier, Swedish union membership has jumped. In the Swedish automobile industry it now approaches 100 percent.[81]

Two lessons for American labor that can be gleaned from Turner's research are that, first, it is possible for unions to play a central role in the corporate world of the future. Second, the complete solution cannot be found in the labor movements of other nations but must arise from the unique characteristics of the local environment.

Are there any domestic examples of unions coping well with the new business revolution? In fact, there are. One is the General Motors NUMMI plant. An important reason for the success of this facility has been the willingness of the United Auto Workers to abandon its traditional categories of worker skills. For example, at NUMMI, instead of the usual two hundred UAW work classifications, there are just three, thus allowing workers to easily move across disciplines.[82]

Another example is Corning, where management, working with the American Flint Glass Workers union, has undertaken a massive shift of its twenty thousand workers to self-managing work teams. The union plays a direct role in the retraining program of company employees, ensuring that its members will not lose their jobs during the transition. The entire process is taking place under the aegis of a statement of philosophy developed by both sides called A Partnership in the Workplace. This statement includes among its tenets the "recognition of the rights of workers to participate in decisions that affect their working lives" and a "work environment free of arbitrary and authoritarian attitudes."[83]

A third, and especially compelling, example of good management/union relations in the United States can be found in the growing number of companies owned by their workers. One of these, as reported in the *New York Times,* is Republic Engineered Steels, of Canton, Ohio. Faced with slumping sales and the prospect of having to lay off 625 of the company's 3,980 workers, Republic's CEO Russell W. Maier turned to the union for help, as it could offer the workplace the discipline and organization he needed.

The two sides searched for answers, among the most effective of which turned out to be soliciting cost-saving ideas from workers. Those with good ideas were temporarily teamed with supervisors to turn notions into reality. The result, after more than one thousand suggestions, was the formation of programs that established enough economies in such areas as water conservation to save more than five hundred of those jobs. Maier and C. William Lynn, president of the United Steelworkers local, had rather symmetrical comments about this result. Said Maier: "You've got to build one ingredient without which you fail ... trust." And Lynn: "We must build mutual trust between one another so we can all focus on long-range job security."[84]

The very idea of such a cozy partnership between management and union is still alien to many labor leaders. And that is the problem.

Again, the virtual corporation is built upon trust and cooperation. Those who cannot accept this new reality risk becoming superfluous. Union leaders face the same challenge as do their long-time antagonists in management. Can they overcome the reactionary elements in their midst and build a new relationship based on co-destiny?

The alternative may be oblivion. As Turner concluded in his research: "It seems to be a particular characteristic of current markets and technologies that managers need more cooperation and problem-solving input from employees at all levels of the firm; and managers can only get this cooperation either by completely excluding unions or by integrating unions into firm decision-making in new ways. . . . Union leaders in the present period, therefore, must be ready to brave internal political obstacles, in the interest of organizational survival, *to move toward a closer engagement with management.*" (Emphasis added.)[85]

To do this may require, as in northern Italy, a new, more decentralized organization in the unions as well, giving locals greater power in negotiating customized agreements with their corporations. In other words, American unions must undergo the same downward shift of power as their corporate management counterparts.

Mere cooperation itself won't be enough. For organized labor to do more than just survive in the future, it must move out ahead to demand more training for workers, more worker empowerment, more union participation in improving productivity and quality, more automation, and more flexibility. This is what will constitute true labor advocacy in the virtual corporation.

## Lost Souls

The most stunning feature of the new work life will be the independence involved. What has been until now the reward for an exceptional few salespeople, researchers, and specialists will increasingly become the rule. Job descriptions will be intentionally vague, rewards often linked to the performance of teams, with the place where the work is to be done sometimes left undefined. Some employees will find that they interact more with suppliers or customers than with their fellow employees, or regularly change bosses, or spend more time with people in far-flung divisions than they do with people in their own building.

All this is going to create some sizable management challenges, more than enough to compensate for the lack of traditional authority. For example, there will be the task of maintaining employee loyalty. When the corporation is almost edgeless, when a worker may operate out of his or her home, or even from a desk at a customer's factory, how does one instill in that employee a sense of belonging? Will a paycheck really be enough? Or will special companywide morale programs have to be created?

Also, as companies become increasingly dependent upon individual employees as the interface to key business partners or customers, what happens when an employee goes on maternity leave or sabbatical or chooses any of the other leave-benefit programs virtual corporations will implement as recruiting tools? Certainly the company can't just shut down the relationship for the interim. Will the rest of the work team, assuming there is one, fill the void? Or will companies have to keep squads of "utility employees" specially trained for temporary fill-in?[86]

But the greatest daily challenge to the workers and the management that supports them will be dealing with the unpredictability of life in the virtual corporation, where perpetual flux will be the rule. If every revolution brings with it the potential for tragedy, then here is where it is most likely to occur.

From one perspective, such fluidity benefits American companies more than their international competitors. As Brit-turned-American Wilf Corrigan of LSI Logic says,

> One of the great advantages of America is that Americans have no memory. The reason I left Europe was because there's such a long memory that you can't initiate change. But Americans have no memory at all. I'm convinced an American workforce can come into work on Monday morning and find the whole production line has changed and by coffee break they're used to the new environment. Americans, unlike, say, the Japanese, are used to change. Most other countries are not. Americans are uniquely adapted to change. Change is the way we can win.[87]

But even in the United States, there is a sizable percentage of people who are change aversive. Many of them migrate to corporations precisely because those institutions have been the most resistant to change. Now a revolution is occurring. What happens to these people, many of them highly successful in the traditional firm? Corrigan shrugs, "We try to find them a position where there's not much change. There will always be a few of those around. But not many."

Solid and steady, among the most admired attributes of the traditional corporation, become negative traits in the virtual corporation. In the process, many individuals who have had trouble fitting the old template will suddenly find themselves in the most amenable of work environments; while, conversely, those who once thrived may discover themselves disoriented, alienated, and overwhelmed by the new work style. It will be one of the sad ironies of the new business revolution created by virtual products that many of the workers and managers who worked so hard to bring that revolution about will find themselves unable, by personality or sensibility, to cross over to the Promised Land. It will be the task of management and labor, working together in a shared task, to help this new group of disenfranchised workers succeed.

# 10

## Spreading the Word

In 1990 and 1991, the most popular advertising campaign in America was for Pepsi-Cola and featured Ray Charles singing "You've got the right one baby, uh-huh."[1] By traditional rights, the success of this campaign should have increased Pepsi's market share against its nemesis, Coca-Cola. After all, Coke's ad campaign, it was generally agreed, had been lackluster.

And yet, according to *Beverage Digest,* Coca-Cola actually gained market share, while Pepsi lost. Something had definitely changed—traditional methods of marketing consumer products seemed to be losing their power. Coke's gains had come from building within its distribution channel, from signing up new restaurant and fast-food chains. Long-term relationships had proven more powerful than catchy advertisements.

During the new business revolution it will be hard to imagine any corporate function undergoing as great a change as that of marketing and its concomitants, advertising and public relations. In fact, it is possible that many of the components of traditional marketing are now all but obsolete. To understand why this is so, we must again look to the legacy of mass production and how it still colors daily business life.

Mass manufacturing had exchanged scope of production for volume. This volume, and the potential profits it offered, drove the creation of immense distribution channels to deliver the fruits of production to the greatest number of customers. As direct sales became increasingly impossible, the great manufacturers turned to alternative tools—local shop owners, catalogs, and newspaper and magazine advertisements—to make the sale.[2]

There was a second challenge as well. The populace, accustomed to custom and semicustom goods, had to be converted to the homogeneity of mass-produced goods. On the whole, this was an easy process; the sheer abundance of goods and their exhilaratingly low prices were in themselves powerful messages.

Still, there was a psychological adaptation that had to be made by the average consumer, and this transition was smoothed by advertisements showing handsome or famous people expressing their satisfaction with the goods in question. Marketing, as a tool for convincing the customer to accept mass production, became increasingly important in the final decades of the nineteenth century and the beginning of the twentieth, as an explosion of new technologies (the kinescope, the automobile, electric appliances, the first plastics), required the customer base not only to buy mass-produced replacements for existing goods but to accept wholly new products without historic precedent.

By the middle of the current century, the cacophony of competing advertisers had grown so loud that it was hard for any one manufacturer to make itself heard above the din. New marketing tools were developed. For example, Fuller and Avon systematized door-to-door selling. The new field of public relations worked to tap into the unbuyable, and thus more perceptually legitimate, editorial sections of publications. Other marketers perfected direct mail vehicles, such as brochures and pitch letters, while still others made use of the telephone and the wonderful new visual technology of television. Still others tried to reach customers by arming distributors and retailers with flyers, coupons, and point-of-sale displays.

This extraordinary outpouring of billboards, neon signs, jingles, commercials, sky-written messages, ballpoint pens bearing slogans, and on and on, is in many ways the signature event of the last one hundred years. Mass marketing has spilled over into every corner of modern life, from the labels on our clothes to the lists of consumer items that now fill much of our literature and art. Yet, in the end, all are just permutations of a single strategy: to shape human beings to act as ready customers for mass-produced goods.

# Turnabout Time

With the arrival of the virtual product, most of this promotional hurly-burly suddenly becomes ineffective. The inverse now becomes the rule. The challenge of the new business era, with its virtual products, is to adapt the product to the consumer, not the consumer to the product.

Such a turnabout will shake the entire marketing profession to its foundations. In the past, by the nature of competition in the mass manufacturing era, most products in time became commodities. Thus, traditional marketing typically began by educating the market on a new product, then quickly settled down to hawking marginal or intangible differences (such as identification with a sports hero) and trying to steal customers from competitors.

In the new business era, in many key markets, the very notion of product differentiation almost becomes less meaningful as the virtual product adapts itself to users. Further, many products will never really become commodities. And, perhaps most important of all, always-a-share customers, those bargain-driven consumers who were the foundation of mass production for a century, will dwindle in number. Instead, customers, deeply committed by the time spent in educating themselves and reconfiguring virtual products to their needs, will be increasingly stable and noninterchangeable among companies. Thus, in a large portion of American business, the impulsive consumer of the postwar era, pulled this way and that by trends and fads, will be gone— and the only sizable consumer population available for capture as new business will include those disaffected with their current suppliers.

Obviously, this won't occur everywhere. Many consumables—for example, athletic shoes and potato chips—will remain commodities. But even these markets will be deeply affected by the new business revolution. For example, manufacturers of packaged goods, until recently among the most profitable of consumer products, are beginning to feel the pinch. As *Forbes* reported in late 1991:

> There are already signs that consumers are starting to balk at the prices of some of their favorite packaged goods and are starting to switch to house brands. The trend so far is only a trickle, but it shows signs of growing rapidly. . . . Once consumers are hooked on price specials and

brand switching, it will be hard to keep them attached to the expensive, nationally advertised brands. . . .

Consumers are inundated with choices and advertising clutter. And, in many product categories, growth is stagnant. Many consumers are also increasingly unwilling to pay up for relatively minor differences in quality or perception of quality.[3]

Advertising testing company Research Systems Corporation estimates that in the last decade advertising persuasiveness fell almost 30 percent among consumers. The response to this by the big packaged goods manufacturers such as Procter & Gamble, Quaker Oats, and Colgate-Palmolive has been a blitz of coupons, discounts, rebates, and product consolidations—all of them threatening to profit margins. And the worst may be yet to come. One Canadian off-brand cola, selling at nearly half the price of Coke and Pepsi, has carved out a huge (20 percent) fraction of the sales of its famous and once invincible competitors.[4] The same thing is happening now in the United States, where, according to Nielsen Marketing Research, store-brand soft drinks are enjoying twice the annual sales growth of overall soda pop sales and experiencing the biggest three-year jump in market share in two decades.[5]

The trend appears common to all industrialized nations. From the United Kingdom, *Management Today* has reported the following:

Manufacturers of packaged goods are under attack from all sides. Traditional methods of branding, such as heavy advertising, are less and less effective as costs soar, television audiences decline, brand loyalty weakens and own label quality improves. . . . Scale no longer provides significant cost advantages to the large manufacturer either; smaller, dedicated suppliers with fewer overheads have emerged. Thus, profit margins have been eroded, and are being further undermined as firms cut prices to bolster declining volumes.[6]

In Germany, consumers are being buried in "a mushrooming of scatterbomb, unsolicited [junk] mail addressed at no specific individual" as companies scramble for workable alternatives to the failing traditional media.[7] And a similar comment has been made by Silicon Valley marketing specialist Regis McKenna:

"Other" owns the leading market share of personal computers, cookies, tires, jeans, beer, and fast foods. Since 1984 American television viewers have been watching "other" more often than the three major networks. Brand names do not hold the lock on consumers they once held. Today, consumers are much more willing to try new things.[8]

All of this underscores the fact that we are entering into an era in which scale, the critical competitive factor in the mass production era, no longer necessarily confers cost advantages. At the same time, variety is becoming cheap. Those features that gave the edge to the large consumer goods makers are disappearing. Profitability is beginning to shift toward customized products that have been tailored (if only, in the case of some packaged goods, by price alone) to their market segments. Thus, even the commodity business will offer few safe harbors from the changes demanded by the new business revolution.

Still, many companies simply haven't recognized the change. Says media marketer Herbert Maneloveg ruefully: "We have decided that the answer to most problems is brand-name awareness without meaningful product differentiation . . . [and] I think we are all suffering because of it."[9]

A business ghetto of price bombing and low-overhead operations will be the fate of many high-flying commodity products of the mass production era. This in turn will have devastating implications for traditional marketing enterprises. Madison Avenue is already suffering, having watched corporate advertising shrink from 60 percent of corporate promotional budgets to just 40 percent—the difference having shifted to direct promotions.[10] "It's murder out there," Metromedia chairman John Kluge told *Forbes*. "The current problems aren't going to be eradicated in six months."[11]

This trough can only be expected to grow deeper as the rise of virtual corporations—with their heavy emphasis on support, training, and customer involvement—further emphasizes the chasm between virtual products and mere commodity goods.

# The Lifelong Customer

Marketing virtual products will be utterly different from marketing commodities—though no less challenging. Here, to find a precedent,

221

one must cast back before the age of mass production and also look at some of the more obscure and anachronistic corners of modern life.

The late-eighteenth- and early-nineteenth-century customers of such furniture makers as Goddard, Phyfe, and Querelle expected to be able to select from a family of basic furniture types (chair, sideboard, highboy, pier table, recamier, and so on), trust the craftsmanship of the builder, and still retain some control over the details and configuration. The same is true now when one orders a custom suit or shoes. An interaction occurs between customer and maker, in which the customer trusts the craftsman's skills and knowledge but still defines the overall style and materials.

With the advent of the virtual product, this trust relationship is restored to large sectors of industry. The goal of virtual corporations is to maximize the binding energy between themselves and their customers. This is done by maximizing customer satisfaction and by enlisting the customer into a co-destiny relationship.

To do this will require enormous investments in time and money—so much so that just as virtual corporations will have to narrow their supplier rolls, so too will they have to reduce their market segments to only the most likely customer candidates. Only in this way can the virtual corporation and its suppliers dedicate enough resources to give the customer the variety he or she requires.

Then, once those customers are on board, virtual corporations will do almost anything to keep them. This means, first and foremost, creating the right products or services, but it also includes special discounts, rewards, a cornucopia of benefits. In fact, almost anything will be cheaper than creating new customers (it is estimated that it costs five times more to create new customers than to keep old ones). This will be the ultimate fulfillment of what Ted Leavitt has defined as the purpose of business: to capture and hold customers.[12] The goal of the virtual corporation will be to capture and hold customers for life.

In the face of all this, what is the likelihood of any traditional marketing campaign unhooking a customer from a virtual corporation? Of course, many firms, to their future dismay, have not yet caught on. Notes Dan Gipple of Britain's O&M Direct: "I am staggered at how many companies are still prepared to sell their product, whether it is a hi-fi or a bottle of spirit, once, without holding onto the information as a vital lead to be approached in the future. In the future it will be about constructing a relationship."[13]

## The Decline of Strategic Marketing

The movement by corporations toward a tighter link with customers is in marked contrast to the most popular marketing techniques of the last few decades. For thirty years the world's capitalist societies have enjoyed an unprecedented period of wealth creation, in the process spinning off not only an amazing number of millionaires and billionaires but also widespread affluence. As has been typical throughout history, such an explosion in wealth has brought with it a number of social distortions, including a focus on form rather than substance that was manifested in conspicuous consumption; greater social respect for financial engineers than for producers of value; and the primacy of image over substance.

During this same period, thanks to the dynamic described in earlier chapters, we have also seen unprecedented advances in technology that revolutionized everything from communications to travel to health care. Coupled with all of this was a demographic force, the baby boom of the 1950s, the largest population bulge in human history—a distortion in the balance of age groups unlikely to be repeated in our lifetimes, one that has played a central role in determining the success and failure of industries throughout the world.

In response to this charged environment, astute marketers began to move away from the more rigid mass-marketing schemes of midcentury toward what might be called strategic marketing. The term is something of a misnomer because it encompasses a number of ideas that probably aren't strategic at all, but nevertheless it does capture some of the complexity of this new idea over the more short-term tactical marketing programs of the past.

Strategic marketing, as we see it, is associated with a number of actions, including market segmentation, product and corporate positioning, marketing warfare (with its hero, Prussian military theorist Carl von Clausewitz), data base marketing, media saturation, inflated claims (as with software "vaporware"), product dramatizations, and sometimes even product falsification (as with the stacked Volvos ad) and event marketing (such as the Apple Macintosh introduction).

Strategic marketing can be seen as an early attempt to come to grips with general prosperity, technological advance, and a demographic aberration as major underlying forces of modern society. Unfortunately for those firms that continue to embrace its philosophy, the world has moved on. With a few exceptions, strategic marketing has limited usefulness to a virtual corporation. Its techniques, with their focus on image and its aggressive intrusion into personal lives, are likely to alienate the same customers the virtual corporation is trying to embrace.

The world *has* changed. After forty years of inflation, we appear to be entering into an extended period that is, if not deflationary, much less inflationary than most of us have known in our lifetimes. "Inflation is a crisis of morality," said General Dorio of the Harvard Business School.[14] Even a less extreme view would have to agree that inflation is product mobile. It rewards speculation over saving, undermines stability, and punishes the prudent.

In the new, less volatile economic environment, much of the transience that has characterized our society in recent years will disappear as well. So too will the era of easy wealth creation. As money regains intrinsic worth, products will be held longer. This in turn will create a greater interest in value and substance. And among the best ways to create value will be quality and service—two processes that do not lend themselves well to traditional or strategic marketing.

This shift toward value is also congruent with the aging of the baby boomer population. Like every generation before it—as it approaches middle age and settles into careers, families, and mortgages—both the hippie first half and the yuppie second half of this generation will become less interested in fad and more in what endures. Market futurist Faith Popcorn has called it "cocooning" and declared it one of the major features of American life in the 1990s. Already one regularly reads of complaints by commodity sellers that they are abandoning the increasingly stable boomers to refocus their efforts on younger generations.

The one radical component of this triad of social forces is technology. But even here an important change is taking place. While the rate of technological progress should continue for years to come, at the practical, everyday level technology is becoming less and less valuable.

For example, jet flight is very valuable to most of us; but as the Concorde situation suggests, supersonic travel is in only marginal demand. Our basic transportation needs are well met by today's high-tech cars—improved versions of the same vehicle will likely produce few more benefits. It is estimated that 90 percent of the features in most computer software packages are never used—suggesting that it is more productive to own twenty easily applicable software programs than one sophisticated but difficult one. Meanwhile, a sizable minority of owners of even as prosaic a product as the VCR has still never learned to program it. And even in medicine, most of the great technological breakthroughs of recent years are relevant only to people in their last months of life.

One can only conclude from all this that one of the most popular marketing tricks of recent years—that of promising customers more and more technology when the incremental benefits are less and less—will lose its appeal. Combine this with a calmer economic environment and an aging population searching for quality and good service, and the era of strategic marketing seems to be nearing its end.

## Value Marketing

What will replace strategic marketing? The marketing of value. The new challenge facing corporate marketing is in convincing customers about corporate and product credibility, quality, service, fairness, and customer satisfaction—the same traits discussed about suppliers in an earlier chapter. The task of the virtual corporation is to develop relationships with customers to enable them to obtain the maximum value from the product they have purchased.

It is not very glitzy stuff. It certainly doesn't evoke Prussian military theorists, Chinese philosophers, or Mongol warlords. If strategic marketing is action packed, then value marketing (like *kaizen*) is positively glacial. It takes years to become creditable, to build a great service infrastructure, and to establish deep and enduring relationships with customers. And the process is very difficult to accelerate—in such a trust-filled milieu, almost any form of hype is seen as betrayal.

225

Value marketing begins by guaranteeing customer satisfaction. Even this initial step is beyond the capabilities of many companies, which have no idea (or sometimes don't even care) if their customers are happy. Those that do try quickly discover that to satisfy a customer one must first find out what that customer expects—which may be utterly different from what the company thought—and then exceed that expectation.

The situation becomes even more complex if the customer proves to have unreasonable expectations (as GM service customers do after watching Mr. Goodwrench commercials) that have to be trimmed.[15] Reigning in customer expectations can be very expensive—Citibank, for example, spent five years and millions of dollars lowering customers' expectations to accept ATMs—but it is often the only alternative to mutiny.[16]

In any event, nothing can happen until the customer's desires have been plumbed. This is where the power of modern information processing can be especially useful. Toyota, with its customer profiling, is only one example of this kind of advanced marketing information gathering. Citicorp, by the end of 1990, was selling market information on 2.3 million users of special ID cards in Chicago, Dallas, Los Angeles, and other major cities. Within a few years it expects to track as many as 50 million households.[17] This information is being used by manufacturers, service companies, and their advertising agencies to target specific customers.

Arbitron and A. C. Nielsen are each spending as much as $125 million to institute nationwide computer networks to gather information directly from supermarket checkout scanners for use by their clients. As Laurel Cutler, vice chairman of ad agency FCB/Leber Katz, told *Forbes,* by using such technology she would soon know the name and address of each person who buys her products, enabling her to target an individual message for each customer.[18] Similarly, Computerized Marketing Technologies, using mailed questionnaires, tracks the product purchases, hobbies, travel habits, and other details of about twenty-five million U.S. households.[19]

Understanding the customer, however, is not enough. It is just as important to measure the current level of customer satisfaction. This is something that so far few companies are willing to attempt, beyond a few desultory questions on warranty cards. Yet this is only the first step. To exceed customer expectations, one must not only design appealing and useful products (a process made easier when, as with a virtual product, that product continuously adapts to the customer) and deliver them in a timely fashion, but maintain a more than satisfactory level of service for that product through its life cycle and that of its descendants.

This long cycle of satisfaction maintenance is the pivot of value marketing. Needless to say, service—defined as all things a company does to enable the customer to derive increased value from the product purchased—is a key factor. If indeed the virtual corporation depends upon keeping customers forever, then great service—service so good that switching to a competitor becomes almost unthinkable—must be central to the business.

Creating good service for virtual products is as difficult as creating the products themselves. It requires the careful administration and interaction of six different factors:[20]

1. *Strategy*—This is the framework that organizes all the other elements of service. The best strategies bring the business into a tight focus upon clearly defined segments. They then attempt to match the company's service strengths with the expectations of customers—even, when necessary, by modifying those expectations.
2. *Leadership*—Leadership makes the strategy work in everyday operations. No company can be an effective service provider unless top management is deeply, almost fanatically, committed to service. This attitude pervades the corporate culture, not only affecting the day-to-day performance of individual employees but easing the acceptance of the long-term sacrifices the company must make to maintain good service.
3. *Personnel*—Because most service is a face-to-face encounter at the front lines where the company meets its customers, the success of corporate service programs depends ultimately upon individual employees. Smart companies will hire employees with the most aptitude for service, train them extensively, then reward them with awards, bonuses, and a just career path.[21]
4. *Design*—Products and services that are not designed from the

start for efficient service will thwart the efforts of a company to achieve that end. In the exemplary service companies, measures taken to ensure a workable design can range from bringing service personnel into the design process, to surveying customers to scale the complexity of design to their competence, to requiring such service equivalents as self-diagnostic software and operating manuals.

5. *Infrastructure*—This includes training programs, inventories of spare parts, extensive customer data bases, and field service personnel. All of this can be very expensive, but, as noted in *Total Customer Service,* "The service leaders we studied charged compensatory prices for the services their infrastructures delivered, grew their infrastructures in line with sales growth, and used advanced technology to cut infrastructure costs and improve performance. Their infrastructures often became bristling barriers to competition."[22]

6. *Measurement*—This is the feedback loop that completes the process begun with strategy. Good measurement systems perpetually test if the other components are still working toward the strategic service goal. Process measurements compare employee performance with the company's standards. Product measurements show whether the work has achieved the desired result. And satisfaction measurements evaluate how satisfied the customer is with the service he or she has received.

Performing all six steps is a huge challenge, but one already being met by the likes of SAS, Federal Express, and Nordstrom. As noted a few years ago in *Total Customer Service* and now more true than ever:

Service standards keep rising. As competitors render better and better service, customers become more demanding. Their expectations grow. When every company's service is shoddy, doing a few things well can earn you a reputation as the customer's savior. But when a competitor emerges from the pack as a service leader, you have to do a lot of things right. Suddenly achieving service leadership costs more and takes longer. It may even be impossible if the competition has too much of a head start. The longer you wait, the harder it is to produce outstanding service.[23]

Good service is just part of a larger goal of value marketing—that of cultivating tight relationships with customers, what Regis McKenna has called relationship marketing. If customers are going to commit time and money to creating tight and enduring bonds with a company they do business with, if they are going to link their economic destiny with, help design products for, or divulge privileged information to that company, they are going to want to know much more about that firm, its philosophy, and its management than they do now.

Ultimately, the customer of the virtual corporation will most resemble the shareholders of that corporation. Both will share a common commitment to the company's long-term success. The industrial customer, even the consumer of expensive goods such as cars or appliances, may have an even greater stake than the shareholder, in that he or she will be less likely to jump to a competitor for only a marginal gain.

One implication of this tightened customer bond will be a change in corporate communications as it takes on some of the traditional trappings of financial relations—reaching out beyond employees to customers with newsletters, financial reports, and other reporting instruments. In the electronics industry we are already seeing the beginning of this with the rise of product user groups, which are not only given unprecedented access to company insider information but even play an important role in the creation of new products. This trend will become more widespread.

For the same reason, managements of virtual corporations will find themselves increasingly attending to a new, and large, constituency of customers. Many top executives will find that they are spending more time with customers and less on internal operations. And, at the apex of the firm, the CEO, already having to deal with a changing role inside the firm, will have the added responsibility of being director of customer relationships. One can find clues to what this will mean in the daily work life of people such as Apple CEO John Sculley—regular visits to corporate customers, speeches in front of user groups, and by-lined customer direct mail pieces.

## Preaching to the Converted

When we look then at marketing in the virtual corporation, we see essentially three tasks: locating new customers, maintaining a two-way flow of information, and maximizing the binding energy between buyer and seller.

Regarding the first task, as virtual corporations become more commonplace the pool of available customers will diminish, reduced in time primarily to the young or the disaffected—that is, those just entering the marketplace or those who are returning to it after being disappointed in a previous business relationship.

A number of noncommodity companies have already begun to target the youth market, with the notion of capturing loyalties for decades to come. In the short term, it also doesn't hurt to tap into the estimated $9 billion in disposable income available from American children aged four to twelve, or the amazing $131 billion in household purchases they are believed to influence.[24]

In the long term, the profit potential from providing a nearly continuous product growth path almost from cradle to grave is enormous. The traditional consumer upgrade strategy required a number of breaks—for example, from an entry-level Chevrolet to an Oldsmobile to a Buick to a Cadillac, and each sold by a different dealer. By contrast, Honda requires of its long-term customers only one jump, from Honda to Acura—and as the power of the brand name continues to diminish, it is likely that even that lone schism will disappear.

Recognizing the importance of capturing consumers early, many companies are starting with the very young. According to *Business Week*, among the firms reaching out for children are A&P stores (with little shopping carts), Delta Airlines (Fantastic Flyer Club), Hyatt Hotels (Camp Hyatt and accompanying newsletters), Sears' Discover credit card (the Extra Credit personal finance education program in *Scholastic* magazines), and Apple Computer (school computer donations and discounts). Says Bill Hodges, Discover's marketing head, "We're looking at 10 years and beyond."[25]

By contrast, reaching the other available consumer population, the disaffected, will require not the creation of loyalty but its restoration. Among the individuals in this second target group will be those who bound themselves tightly to a company, only to be left disappointed, angry, even feeling betrayed—because of poor service, being ignored, or encountering a dead end in what was supposed to be a continuous product migration path. These potential customers will not be reachable through traditional advertising or promotional techniques. Cynical and wary, the one sure way they can be touched is by word of mouth from people they trust and respect. Companies that want this business will be most able to influence the choice by facilitating the travel of this word of mouth—such as by republishing analysts' reports or through the types of ads that say, "Ask your friend who owns a Volvo."

Maintaining a two-way flow of information of sufficient breadth to carry not only sales figures but ideas, opinions, and desires of both the virtual corporation and its customers will require massive amounts of information technology and a reevaluation of existing marketing channels. One way to do this is through market segmentation. Wrote Marian B. Wood and Evelyn Ehrlich of Business Marketing Group:

> Once the exclusive domain of the consumer crowd, market segmentation is now an important tool for a growing number of industrial and business-to-business firms. . . . By breaking down the overall market for their products into smaller segments, companies can tailor sales and promotional strategies to the needs of specific customer groups, in addition to developing new products that meet those needs. . . .
>
> For smaller companies, segmentation can be a way of outflanking large competitors. By identifying markets that are either particularly profitable on the one hand, or underserved on the other, an agile company can design segment-specific strategies that will establish it as the market leader in their customer's minds.[26]

Wood and Ehrlich list as examples the Hartford Insurance Group, which expanded its overseas coverage after learning that its fabricated metals clients were entering export markets; Philadelphia's Liberty Bank, which increased market share after spotting the overlooked market of small business customers; and Minnesota's Marvin Windows, which developed custom marketing approaches to the most profitable commercial window market segments and grew from eighth to third in its industry.[27] (Note, however, that the success lay not just in the segmentation but in following that segmentation with products that satisfied those customers.)

Industrial marketing consultant Harold J. Novick has also pointed out that with market segmentation must also come sales channel segmentation. If the sales force is not adequate or competent enough to reach the newly segmented market niches, then the entire exercise will be futile—the customer will not get the services needed to support the product.[28]

In the virtual corporation, market segmentation becomes market atomization. Niches, especially in the industrial market, may contain only one customer—as with Intel, which has an IBM products division. But if a company is not fully prepared for this sudden explosion in business units, it will be unable to manage them. Even for prepared companies, this level of segmentation will demand some ruthless pruning of the customer base. Customers who are unwilling to be trained, not forthcoming with necessary information, or not willing to participate in a shared future must be quickly culled out or they will become a heavy financial burden. For this sophisticated measurement, systems must be in place from the start to help make the streamlining fast and accurate.

For those customers who are desirable, the virtual corporation will have to move fast to emplace a multiplicity of custom communications connections—print, electronic, and human—to gather information, begin training, modify product designs, speed manufacture and delivery, offer rewards, and in every way possible quickly bind that customer to the relationship.

There are a growing number of companies, both consumer and industrial, that are beginning to do just that. Richard Winger and David Edelman of the Boston Consulting Group have identified three steps in what they call "the art of selling to a segment of one." The first is installing the requisite information processing equipment to maintain extensive customer files. The second is finding out how to tailor a service or product to individual consumers. They cite French cosmetics maker Yves Rocher, which maintains customer mail-order purchase histories in order to regularly generate new custom order forms, and the Four Seasons Hotel chain, where a phone call to the lobby from a room immediately fills a nearby computer screen with extensive information on that guest's special needs.[29]

The third step goes beyond this to a regular and personal communication with the consumer. An example is Merrill Lynch, which successfully introduced a new investment product by mailing its customers a personalized portfolio analysis that showed what the return would be with an investment in that new product. Other types of personal communications they list include selective binding to create custom magazines, videotex, point-of-purchase communications, and targeted co-op mailers. To this, one might add computer bulletin boards, interactive voice mail, newsletters, users' group meetings, and using the virtual product itself as a means of communication and information gathering.

Winger and Edelman's conclusion is similar to that of the other voices in this chapter: "A successful segment-of-one marketing strategy requires a broad rethinking of the value a company provides to its customers—and significant investments in service and information. Competitive advantage will tilt to those companies that can satisfy individual customers' needs."[30]

## Distributor Satisfaction

Almost everything that has been said about binding the customer also describes the relationship that must be developed between virtual corporations and distributors and retailers. In fact, companies must abandon the old image of distributors as mere contractors and of retailers as disloyal shopkeepers and actually try to bind both even tighter than the final customers.

Distributors and retailers actually are surrogate employees. As such, their sense of trust and co-destiny must be extremely high. After all, they will serve as conduits for much of the ongoing linkage taking place between the virtual corporation and its long-term customer—in McKenna's phrase, they are the "communications mediums" to customers. It is their information gathering equipment that is the first line of contact with the customer base. And, in the case of retailers, it is their employees who will represent much of the human contact with the virtual corporation's customers. A virtual corporation might go to superhuman lengths to satisfy its customers, only to have a decade of work go down the drain thanks to a nasty sales clerk or a lazy distributor.

This suggests that the competition to line up long-term distribution channel participants for virtual corporations may turn out to be even hotter than that of signing on loyal customers. Some of the most successful corporations in America have very tightly bound distributor networks. Author Milind M. Lele, in studying customer-oriented companies, noted that Caterpillar, John Deere, Century 21, Maytag, and Mercedes-Benz—all premier companies in their markets—have legendary relationships with their franchisees, distributors, or retailers.

Lele came to several conclusions about these intermediaries: satisfaction with the dealer and overall satisfaction go hand-in-hand; dealers determine customers' attitudes and expectations; dealers affect product performance; and dealers influence customer feedback and restitution. "Intermediaries," Lele wrote, "can make or break a firm's efforts to establish a franchise with its customers and are thus a critical element in ensuring overall customer satisfaction."[31]

For example, many Deere and Caterpillar dealerships have been held by the same family for three or more generations. Lele reports that Caterpillar even has a school to help the children of dealers get into the business. Yet, despite this chumminess, both companies are very demanding of their dealers. For example, they must maintain a sufficient level of net worth. Store layouts are controlled and the dress of dealer personnel is tightly defined. Training of dealership personnel matches that of the factory workers. The dealers themselves are expected to become important players in the community. Wrote Lele: "The local Deere dealer is usually held in high regard by his or her peers: Often, he or she sits on the board of the local bank, is active in civic affairs, and is usually a community leader of some stature. This has definite benefits for Deere in that the dealer's reputation for honesty and fair dealing reinforces Deere's image."[32]

The quality of these intermediaries and the high levels of customer satisfaction they support (Deere is consistently rated highest in its industry) have probably saved both Deere and its dealers in recent years.

The biggest threat to John Deere came in 1974. The company was late to market with a four-wheel drive tractor—only to have the $60,000 machines suffer a high breakdown rate. It was the dealers who saved the company, first by providing Deere with complete and accurate feedback on the problem—data Deere's management trusted because of the close relationship with the dealers—then by holding onto farmers' loyalty until Deere could redesign the faulty parts. "Essentially, they said to the customer, 'We know they've had some problems, but don't worry about it. I'm here and Deere's here to stand behind the equipment and fix it, regardless.'"[33]

What do John Deere dealers get as a reward? The company provided the highest returns on investment in the industry. But more than that, it returned loyalty for loyalty. When the industry was hit by a devastating slump in the early 1980s, most of Deere's major competitors abandoned their dealerships. Deere managed to retain most of its dealers. Wrote the *Wall Street Transcript* in 1983, "Deere has gone around and actually to the detriment of their balance sheet given support to their dealers. They're the only people who are going to end up with a healthy dealer network. Ten to twelve percent of the farm equipment dealers in 1982 went out of business, but the number of Deere dealers that went out of business was minuscule."[34]

Such interdependence breeds respect. John Deere's dealers are not just the company's representatives to the customer, they are its advocates. They do not merely gather market information but interpret it, participating in the manufacturer's future success (and enlisting customers to do the same). It is just this kind of distribution channel interrelationship that will be the key to successful marketing by the virtual corporation.

## An Inside Pitch

As we enter an era of tightly bound distributors and customers, a diminishing population of potential new customers, and alternative means of marketing communications, what is the role of the traditional arts of advertising and public relations?

As noted, modern advertising was created by the conjunction of mass production and mass media. Its task was to convert individual tastes into a desire for the generic fruits of mass production. That it often accomplished this through appeals to individuality, snobbishness, and uniqueness is a testament to its cleverness. Public relations assumed a similar role, only its target was the press and its task was to convince reporters to write favorably about the company's products or business. Even with the rise of niche-oriented trade magazines, advertising and public relations retained this philosophy, rooted in individual products and their appeal to potential customers.

It is hard to imagine how these practices can continue as effectively in the future. Certainly there will always be some need for clever mass promotion of commodity goods. But the number of these products is diminishing with the rise of virtual corporations.

Among virtual corporations, far more important than finding new customers will be serving the ones they already have. Advertising and public relations departments are likely to see reduced budgets as money flows toward the maintenance of customer satisfaction and away from the job of creating new prospects. For the virtual corporation, money spent on an easier-to-open shipping box or a better-written manual will probably be more valuable than that spent on a four-color ad in a national magazine.

The advertising functions that do survive the new business revolution will be those that have adapted to the changed business reality. There are already some clues as to what these changes will mean. In the ASIC business, for example, companies such as LSI Logic seldom advertise individual products. After all, such ads would be worthless, as each product is customized. Instead, these companies use advertising to tout their capabilities, to explain what they can do for potential customers. The companies then often supplement this with personal contacts at the highest levels. Says LSI Logic's Wilf Corrigan: "In this relationship, the ties that bind are not made exclusively at the purchasing agent-salesman level. The relationship I'm describing involves an interface of top technologists as well as top corporate management. There is too much at stake to devote less firepower to the relationship."[35]

Public relations' new role will be educating (and, if possible, influencing) those in the know. Regis McKenna has long argued that as products grow more complex and competition more intense, both industrial customers and consumers increasingly look to key opinion makers for advice. In any given industry, these opinion makers, respected for their knowledge and objectivity, rarely number more than a handful. Their judgments can make or break a product, rendering worthless the millions spent on product advertising. As an example, he points to the IBM PCjr, which died despite an expensive advertising campaign because the key analysts, columnists, and technologists in the personal computer industry gave it poor reviews.[36]

McKenna's argument is echoed in *Business Week*, which, in calling the 1990s the decade of value marketing, suggested that companies "use your advertising to provide the kind of detailed information today's sophisticated consumer demands" as well as offer "frequent buyer plans, 800 numbers and membership clubs [that] can help bind the consumer to your product or service."[37]

Successful advertising and public relations in the future will enhance the binding energy of the customer to the virtual corporation. While this may include traditional high-concept ads in mass media venues, it is hard to imagine they will ever again have the same importance in noncommodity marketing campaigns. Rather, they may serve as an occasional reinforcement or, with mass market public relations, be used to reach shareholders and government policy makers.

The real marketing communications battles of the future will take place among the niche magazines, cable television stations, computer networks, and industry newsletters. The goal will no longer be to capture a share of mind but to deliver peace of mind, an assurance that the virtual corporation you have joined up with is well managed and will take good care of you. The job of advertising (and direct mail) will be to make sure every customer regularly hears that message and believes it. The task of public relations will be to make sure that the opinion makers those customers rely upon have received sufficient information to make considered and accurate judgments. To do this, marketing communications people, especially those in agencies, will have to become experts themselves in the virtual products they promote—except for certain commodity products, the age of the star copywriter and "You've got the right one baby, uh-huh" is coming to an to end.

One more clue to the future of marketing communications for noncommodity products can be found in the writings of Guy Kawasaki. A former Apple executive, Kawasaki argues that in a world of informed consumers who see through most advertising hype, the sole remaining course for marketing is passionate sincerity—"evangelism," as he calls it:

> Evangelism is the process of convincing people to believe in your product or idea as much as you do. It means selling your dream by using fervor, zeal, guts and cunning.
>
> In contrast with the old-fashioned concept of closing a deal, evangelism means showing others why they should dream your dream. . . . Evangelism is the process of spreading a cause.[38]

Kawasaki's idea of a shared dream may be the best description yet of the changed relationships that will emerge from the new business

**11**

········································································

# Toward a
# Revitalized Economy

*Virtual corporations cannot survive unless they operate
within a supportive economic environment.*

How can virtual corporations gather all the information they need
or locate trained workers, suppliers, and customers if they are
immersed in a sluggish, retrograde business landscape, if the
transfer of data and material is across an archaic infrastructure, or if
the general population is poorly educated, alienated, and intellectually
inflexible?

Without a nurturing social, political, and commercial environment,
virtual corporations will sprout in response to international competition, then wither in a hostile climate. History is replete with examples
of countries that eviscerated emerging industrial transformations and
in the process fell behind their more adaptive counterparts. The most
famous of these transformations was the Industrial Revolution, which,
though its core technologies were within the reach of dozens of countries, found fertile ground only in Great Britain and the United States.
Much of the rest of the world has spent two centuries recovering from
its mistake.

The new revolution in business is likely to present us with the
same kind of jarring historical discontinuity. And, like its great eigh-

239

teenth-century antecedent, this revolution will not be merely a commercial transformation but a sociopolitical one as well. The revised notions of work, employer–employee relations, and even knowledge engendered by this revolution may ultimately lead to a different way of looking at the world. If the driving ideas of the Industrial Revolution were energy, specialization, and replaceability, in this new era they will be time, learning, and adaptability.

Hints of what is to come already exist. We stand in awe as a few American and Japanese companies seem to accelerate before our eyes into another dimension of productivity. New products appear and are then improved or replaced at a seemingly impossible pace that grows yet faster by the year. Slower competitors are soon overwhelmed by this flood, their products quickly appearing anachronistic. These new corporations seem to do it all better. Sun Microsystems and Dell Computer are turning the computer industry upside down. The Lexus sedan in its first year not only came to market quicker and cheaper than its European luxury counterparts, but it also offers greater performance and the highest quality ratings in the history of the automobile industry.

We are amazed because this has never happened before in our time. For decades, most business change has been incremental, not momentous. But it has happened. Our amazement is similar to that of the Belgian weaver of 1790 visiting a British water-powered loom, or of a Frenchman touring the Remington Arms model plant in Paris in 1850. The ground beneath us has begun to move, and we sense that when the shaking stops, the very topography of our lives will be utterly changed.

## Close to Home

Here in the United States, this sense of distortion and confusion, mixed with considerable fear, has become an uncomfortable part of our daily lives. Everywhere there is a disquieting sense of decay—in government, within boardrooms, on shop floors.

The United States remains the world's richest and most powerful nation, but the news otherwise is not good. Even our great wealth seems tenuous when one considers that we are also the world's largest debtor, we run huge budget and trade deficits, we are losing our manufacturing base, and we face a litany of social ills.

Our schools are in disrepair, many school districts are going bankrupt, the curriculum is being torn apart by factionalism, test scores continue to drop—and American schoolchildren are falling ever further behind their counterparts in other developed (and even some underdeveloped) nations. On an international examination given to nine- and thirteen-year-olds, American schoolchildren rank at the bottom in every area but leisure time.[1]

One by one our industries are losing competitiveness and market share to industries of other nations. Our government seems more concerned with lifelong job security for politicians and spending money it doesn't have than in enhancing the economic prosperity of the citizenry. Our manufacturing sector often insults consumers with shoddy products and workers with unearned executive compensation—and then blames its woes on foreign competition. By the same token, workers are frequently unmotivated and selfish compared with their foreign counterparts; and consumers have in the past replaced good sense and security with almost pathological acquisitiveness.

Meanwhile, our major cities, once the jewels of our culture, have become violent, ungovernable places perpetually teetering on bankruptcy. In some parts of the country, there is ongoing debate whether there should even be a common language. Litigation has become one of our few growth industries, leaving other, more productive institutions and professions cowering in its shadow. Our most talented people now spurn government careers, a dangerous trend in a representative democracy.

Proven techniques for solving these problems are suddenly ineffectual. Keynesian monetary policy fails in a world of multinationals and the instantaneous global transfer of currency. Tax cuts spike purchasing but, unmatched by commensurate spending cuts, also increase the debt. Meanwhile, American competitiveness continues to slip, as does the real income of its citizens. Worse, this democratic society seems to have become a new kind of house divided against itself, in which the upper class enjoys prosperity while the middle class struggles to not lose ground and a recalcitrant underclass drifts off into a multigenerational descent into crime, drug addiction, and government dependency. Writes economist Paul Krugman:

Although some people became fabulously rich, and a sizable fraction of the population achieved unprecedented affluence, the typical American family and the typical American worker earned little if any more in real terms in 1988 than they did in the late 1970s. Indeed, for the median American worker there has been no increase in take-home pay since the first inauguration of Richard Nixon. And for Americans in the bottom fifth of the income distribution the 1980s [were] little short of nightmarish, with real incomes dropping, the fraction of the population in poverty rising, and homelessness soaring.[2]

Although there are solutions to these problems, they will not be easy. What is clear is that they must in part be based on an economy capable of producing manufactured goods competitive in world markets as well as agricultural products and services.

We must produce goods in order to create jobs in both the industrial and service segments. So much of the service segment is both directly and indirectly linked to industry that it cannot prosper without an industrial infrastructure. Evidence of this is all around us. As the *New York Times* reported at the beginning of 1991:

In the 1980's, when services added a stunning 21 million jobs and employed almost four out of five workers, Americans debated whether service jobs were good jobs or bad jobs, but basically took the steady growth of services for granted.

No more. Except for health care, the services are in the throes of a pervasive shake-up very much like the one that racked smokestack manufacturers a decade ago. . . . [Economists] expect job growth in the 90's to be the slowest since the 1950's . . . and they predict that job security—the comfortable expectation of being able to settle down somewhere for life, at least by middle age—may be gone for good.[3]

We need products to export in order to earn the foreign exchange so we can afford to buy the goods and services we cannot effectively produce in our country. We need manufacturing to fuel the economy in order to provide jobs and the dignity that comes with them for many of our citizens.

Many of the current problems of the United States, especially those in business, arise from a nation struggling and stumbling through the no-man's-land between one industrial era and the next. Some of our businesses are already climbing into the twenty-first century, while many are still trapped in the labor wars of the 1930s or in management theories of the 1950s. The laggards, despite heroic efforts, are losing ground to new domestic and international competitors because they've failed to notice that the game has not only changed but has moved to a different field. Unless they adopt a different approach to their business, they will be lost and will not be able to be a part of the solution to the problem.

# The Arrogance of Success

Mass production was probably the single most important factor in making the United States the world's most powerful economy. The processes conceived by Henry Ford and Frederick Taylor became the dynamos of the world's largest manufacturing corporations. Riding in their wake came the great banks, retail chains, and a host of professional service suppliers, including doctors, lawyers, and accountants.

The spirit of the industrial corporation was perhaps best expressed by GM president "Engine" Charlie Wilson in a hearing before Congress in 1953, when he said, "For years I thought that what was good for our country was good for General Motors, and vice versa."[4] Those words still echo today through the empty and failed factories that stand as monuments to our industrial shortsightedness.

When mass production was at its apex, the captains of industry were promoted to be generals of the economy. Success having validated their brilliance, they focused more upon serving their own needs than on serving those of their customers or of society. They epitomized forward thinking—and did so long past the day when, in Benjamin Coriat's words, it had become necessary to "think in reverse." Instead of taking guidance from their customers and turning that wisdom into strategy, corporations instead predictably chose to mold the customer base to meet their needs. The task became one of shaping demand to meet the needs of mass production rather than sculpting production to satisfy customer needs and tastes.

Communities and governments were expected to yield to industry in this Grand Scheme. They were expected to support and pay the

243

social costs of the corporation. They were the ones responsible for the blackened skies of Pittsburgh and the polluted waters of the Cuyahoga River. They were to pay the medical costs of injured or health-broken employees. They were to train the interchangeable worker. And most of all, they were to devise laws favorable to the industrial interests.

The logic behind this was elegant in its simplicity. What was good for General Motors was also good for US Steel, Du Pont, Standard Oil, and all the rest. And what was good for those companies was good for their employees, who, because they were gainfully employed, now had money to feed and clothe their families and serve as the engine for economic growth. The worker would purchase the output of the factories, buy food from the farmer, and provide taxes to support the government. It was to be an economic perpetual motion machine.

Mass production was based on Frederick Taylor's one best way and Henry Ford's one best car, the black Model T. According to this worldview it was inevitable that Eli Whitney's interchangeable parts would presage the need for the interchangeable worker. Work was simplified to the point where almost anyone could be trained to perform repetitive tasks effectively. The education system adjusted its training programs to meet the needs of the market.

Management trained the worker to take orders rather than think. The one best way demanded workers who were only allowed to do one single thing. Work become more rote and management more rigid. Labor responded to this rigidity with rigidity of its own. Unions grew in power, and with that came even more stultifying work rules. Now industry found itself with workers who were allowed to operate machines but were not allowed to fix them. Or workers who were not allowed to carry materials from one place to another. Or maintenance people who could not perform simple plumbing and electrical jobs because they did not belong to the right union.

Needless to say, the philosophy of one best way, of one best car, led to production facilities that were as rigid as the workers and the managements. These single-purpose production facilities could do only one best thing. As a result there was one best marketing philosophy as well. The reasoning was inevitable: if the factory could produce only a limited variety of products, why not shape the markets to demand the products the factory could produce?

Products became ever more standardized and ever less differentiated. Faced with this, companies began to focus on low price as the way of capturing customers. They became rigidly attuned to driving costs down. Cost, not quality, became the key to market share. Many company managements came to believe that low cost was in fact contrary to high quality. Sacrificed as well was customer responsiveness. Since the rigid systems could not deliver it, it was not important.

The notion of interchangeable workers and parts dovetailed neatly with that of interchangeable suppliers and replaceable customers. Suppliers existed to serve on the terms most favorable to the industrial customer. And if the supplier balked at those terms there was always another one to turn to. Therefore, most attention was focused on price as opposed to quality and relationships. There was little trust on either side of the bargaining table and little interest in attempting to help one another. Secrets were important, and information of value to the other side was tightly guarded.

As for the marketplace, an attitude of caveat emptor was just fine with the corporations. If the consumer was responsible for buying a poor-quality product, there was little need to worry about the quality of what was produced. This was especially true after World War II, when customers were numerous and supply was restricted. In a sellers' market, unhappy customers were acceptable. They had no where else to go.

As myopic as it may seem to contemporary eyes, this business model not only worked but worked brilliantly. The U.S. industrial engine managed to create the wealthiest society in history while simultaneously overwhelming the two great totalitarian threats of the age, communism and fascism.

But hubris accompanied the acquisition of this power. And when cracks began to appear in the monolith of American manufacturing, the great industrial leaders quickly blamed others for their problems. The villains included indolent and ignorant labor, the high cost of capital, trade restrictions, tax policies, litigation and product liability, predatory foreign competition, onerous environmental regulation, a poorly educated work force, among many others. Certainly all had contributed to the situation. Yet ignored in all this were the flaws in the internal mechanisms of the corporations themselves.

When it became obvious that America's industrial engine had begun to sputter, many economists and leaders, refusing to admit to a more systemic failure, clutched for the straw of the so-called postindustrial economy. This theory not only provided a convenient explanation for the country's loss of manufacturing prowess, but even made it seem an advantage. After all, what could be cleaner and nicer than a place where brain work had replaced turbines and smokestacks? Not surprisingly, it was a vision that corresponded perfectly with the personal fantasies of many academicians and corporate staff members. In the new postindustrial society, the problems of grimy manufacturers would mean little if everyone could find cleaner work. Hollow manufacturing corporations could be created that served as marketing organizations for the mass-produced products of Asia. Service businesses would flourish and many would find work as doctors, lawyers, accountants, bankers, stewardesses, retail clerks, or hamburger-flippers in fast-food restaurants.

Of course, the populace was not spending a lot of time worrying about the situation. It was enjoying the great wealth created by the production machine. Since people could get jobs without finishing school, there was little need to finish. Since government would take care of them in their old age, there was little need to save. Since many could live well enough on welfare, there was less need to work. The government was at work creating a hollow economic structure to match the hollowed-out factory.

Through all this, government stood by, paralyzed by indecision. One faction argued that services would propel the country to the next level of prosperity. From the other side came the argument that industry could solve its own problems if left alone or if foreign competitors were held at bay for a few years.

Meanwhile, business did little to help its own cause. In a babel-like atmosphere, some leaders argued for protectionism. Others begged to be freed from the shackles of unfair labor, unfair laws, and unfair litigation. High-tech entrepreneurs announced that all they wanted was for government to create a level playing field and then get out of the way. About the only subject on which all seemed to agree was that the government should not become involved in industrial policy. Unfortunately, for good or bad, this was the one activity in which federal and state governments in the United States had been involved for years.

It had never been billed as such, but nevertheless, the U.S. tax code was the de facto industrial policy of the United States. It determined what type of investments would be most attractive at a given time: real estate sometimes, capital equipment at other times. By making interest deductible and taxing dividends, it sparked a leveraged buyout spree. By permitting tax-free institutions to avoid paying tax on investments in stocks, the code did a great deal to influence the trading mentality in the stock market and management's focus on short-term results.

In fact, the tax code was only the most sweeping form of the so-called nonexistent industrial policy. Everywhere in government one could find industrial policy being made. Product liability laws determined the production of everything from pharmaceuticals to small airplanes. Industrial revenue bonds and tax incentives determined the sites of factories. Defense industry expenditures determined how many engineers would be available to work on consumer products. Educational standards determined what future employee talents would be available for industry. Immigration laws determined which job classifications would be filled and which would go begging. Rigid antitrust laws designed to solve the problems of 1910 determined which threatened industries would be allowed to die before they could form research cooperatives.

What was especially pernicious about this denial of the existence of the industrial policy was that what was created happened in piecemeal fashion, without coordination and with no common goals. If we were to have a national industrial policy, then it should at least be a coherent one with focused national initiatives in key areas.

Business executives who did try to remain competitive found themselves perpetually reeling from the latest quixotic decision from Washington. No sooner would an American company show some success in a foreign country than the U.S. government would indulge in some fiscal and monetary policy that would force the dollar sky-high and drive its business venture into the red. That night, the company president would turn on the TV and watch the U.S. president declare the strong dollar to be a policy victory.

Over time, this combination of social ills, ineffective government, and business myopia began to degrade the country's commercial institutions. Christopher Hill characterized manufacturing's problem as one of "continuous incremental degradation," saying also that "they don't make them like they used to." The service industry fared no better, its failures best captured in the poignant question on the cover of *Time* magazine: "Why is service so bad?"[5]

247

As it became harder and harder to create wealth in a society having trouble manufacturing competitively and producing quality services, enterprising individuals decided to redistribute wealth through manipulation. Financiers did this with leveraged buyouts, bankers created the house of cards from which the savings and loan crisis grew, and plaintiffs' lawyers reshaped the law so that seemingly everything became a tort. Government as well focused on wealth redistribution rather than wealth creation. To avoid forcing society to make the tough choices, it ran up large deficits and used inflation as a way to tax the public without being blamed.

Now, in the 1990s, the disturbing truth has become undeniable. For twenty years, the annual growth in per capita income has been at a historic low. Unless something is done soon, for the first time this century, this generation will live less well than its parents did.

What had been obscured and overlooked by all this was the opportunity to create a new type of added-value production—one that values the needs of customers, suppliers, and employees more and the corporate hierarchy less. It would be an industrial model built on cooperation among business, labor, and government, not destructive confrontation. What was needed was a "thinking in reverse."

## A Different Story

The Japanese didn't make the same mistakes as did the United States. Having rebuilt a war-decimated economy in the 1950s, they were not encumbered by the burden of past success. Their Golden Age would occur in the future—and the nation, aided by a hermetic and homogeneous culture that could be organized around a common goal, set out to achieve it. Not locked into old patterns and at the same time searching for a way to differentiate themselves from their established American and European counterparts, the Japanese found quality—learned, ironically, from Americans such as Deming and Juran. As a result, they were the first to emplace a new industrial paradigm built on constructive cooperation (rather than destructive confrontation) among business, labor, and government. They adopted a new theme: "What is good for the customer is good for business."

Japan was heavily rewarded for being the first to make this crucial step toward the creation of virtual corporations, jumping into the ranks of the world's richest nations in just twenty years. But hubris is an international disease, and Japan for its arrogance is now too beginning to feel its sting.

While it is clear the Japanese have done a better job of paying attention to the needs of the customer in many areas (e.g., automobiles and consumer electronics), it is equally clear they have often exhibited reckless disregard for the needs of the economies in which those customers live. Their inability to share their success with other industrialized nations has done much to destroy the markets on which they depend. Now they are becoming victims of restrictive import policies in Europe and similar forms of Japan-bashing in numerous other countries of the world.

Furthermore, Japanese corporations have exhibited a singular lack of trust in their non-Japanese managers. This, according to research by McKinsey & Company, creates a vicious cycle in which foreign managers working for Japanese companies perceive limited career opportunities and leave the firm—to be replaced by Japanese managers. This in turn further reduces the perceived opportunities for foreign managers. As a result, "the Japanese organization is often left with insufficient local talent."[6]

Finally, with other Asian countries having driven the Japanese from the low added-value commodity markets, it becomes apparent that Japan's future depends on selling advanced entertainment systems, high-end color TVs, luxury cars, advanced color copying systems, camcorders, and feature-loaded 35-mm cameras to prosperous economies—precisely those that, pounded by Japanese indifference, are beginning to suffer economic distress and lash back. This would seem to be as self-destructive a long-term strategy as the United States had in the 1950s, yet it is being assiduously pursued not only by Japan but the hungry developing countries of Asia.

## A Domestic Revolution

It must be recognized that there will be no great masterstroke or earth-shattering new invention that will revivify our industry and restore vibrancy to our economy. To succeed, a transformation must take place

249

in almost every office of almost every company in almost every industry in the nation. Furthermore, it must also be recognized that before wealth can be reinvested, redistributed, or spent, it must be created.

The only way a large, advanced society can do this is to create high-quality, value-added products and services. We cannot maintain, much less improve, our standard of living by building low-cost, undifferentiated commodities and pitting our highly paid labor against that of the developing countries of the world. A nation engaged in commodity production will soon find that its wages must on some adjusted basis be equal to those of developing countries. If an American auto worker hopes over the long term to earn five times as much as his Asian counterpart, he or she must be roughly five times as productive.

There are secondary effects as well. If a large segment of the population is paid as if it lived in the Third World, then managers, doctors, lawyers, accountants, and others will soon find they are living in a society that cannot afford them unless they work for less. If the people working in the hospitality industry hope to be able to fill rooms in deluxe hotels and tables in restaurants, they must sell at prices that are within the reach of the buying public. This of course links the wages in the hospitality and travel industries to the incomes of other members of society. This is already happening around us, as doctors' incomes have been squeezed by government programs and as members of other service professions find it harder to find clients to pay for their services.

The new business revolution created by virtual products is crucial because it is currently the best way to add value to an economy. Virtual corporations will have faster and cheaper product development, their products will be more efficient and of higher quality, their grip on market share will be more secure, and their customers will be more loyal. An economy built upon virtual corporations will be, despite its continuous internal flux, an extraordinarily stable and enduring structure. Its citizenry will be highly trained, its institutions strengthened by bonds of long-term relationships, its productivity unsurpassed, and its products and services competitive in the world market.

But the question remains: Which of the world's economies will have the courage to lead the way into the new business revolution? After all, it will come with considerable cost, sometimes to what are now our best workers and managers; its core philosophy is considered counterintuitive by many contemporary business leaders; and it will demand a level of trust higher than most people currently consider safe.

It seems unlikely that no nation will cross over into the new industrial era. That would be unprecedented in the history of both technology and commerce. Some nations, perhaps even some currently uncompetitive ones, will make the leap. And, as these economies accelerate away, the rest will be quickly left behind, victims of their own shortsightedness.

What then would it take to make the United States the leader of this change? The answer is: whatever it takes to best nurture and grow virtual corporations.

The virtual corporation in turn will do best in a society that is highly educated with a high level of technical and computer-related skills. It will most likely succeed in a country that possesses leading technology in fields related to consumer needs. It will be best able to function in a country with an excellent communications and transportation infrastructure. It will work best in an environment where there is teamwork between business and government. It will function best in a society that is less confrontational. It has its best chance to survive in an environment dedicated to long-term goals and constant improvement. Its interests are best served by a society that saves more and consumes less. And it will proliferate most widely in an environment that rewards business contributions that create jobs and wealth rather than merely rearrange them.

## Social *Kaizen*

This book has so far dealt mainly with what corporations must do if they wish to remain competitive. But these corporate transformations will be heavily dependent upon the level of support they receive from society. Corporations will not be able to cope with the competitive challenges of the twenty-first century by themselves. It is therefore important to inquire as to what government and society should do to support this movement.

The first thing to understand is that the transition is not going to occur overnight. It will go on for decades. Continuing support will be required from social and government programs during this interval. Yet, at the same time, any abrupt or dramatic change in government and social programs is more likely to hinder than help.

Our society has become so complex, with so many subtle and conflicting forces, that it is almost impossible to predict a positive outcome for any one program, no matter how well intended. For example, welfare programs aimed at making individuals less dependent have only created, according to a number of observers, more dependency. The war on drugs has made little progress despite thousands of arrests and billions of dollars spent. Industrial tax credits have seldom led to more productive long-term output, but rather, in many cases, have created only bookkeeping profits. Protection of the steel industry has not led to the expected increased investment in the steel business by those who cried loudest for it.

In the future, the best government and social programs will be those that operate under the rules of *kaizen*—taking small gradual steps toward a goal and evaluating the progress before going further. If the program isn't working, then stop—either abandon it or fine-tune it so that it will succeed.

It is not clear that any legislative body could ever restrain itself to the degree required by the principles of *kaizen*. Unfortunately, the consequences of not doing so will be to predestine the nation to constant flip-flops in direction, good ideas becoming expensive mistakes to be dealt with by reforms that start the cycle over again.

In order for this process to work, both major political parties must agree on some broad policy objectives and work toward these goals through successive changes in administration. This implies that the United States must have an industrial policy, the goals of which will obviously not remain completely static over the years—but neither can they be allowed to change dramatically with every election.

## Industrial Policy

Here we tread on dangerous ground. The phrase "industrial policy" has become so burdened with emotional baggage, so identified with statist solutions to commercial problems, that it is almost impossible to discuss it objectively.

As noted earlier, the United States already has an industrialized policy—a cobbling of laws, regulations, Commerce Department rulings, tax codes, tariffs, subsidies, and embargo lists. The problem is that it is bad policy. It is punitive when it should provide incentives, obsessed with minutiae when it needs to be strategic, and rewards established industries with good lobbyists rather than giving emerging industries a vital stake in the future.

If the United States is to have an industrial policy, then it should be de jure, not de facto; and it should be dedicated to supporting the virtual corporations of the future, not to protecting an anachronistic past. There is a way to do this, one that divorces itself from the chaos of daily politics yet is answerable to the long-term needs of the country. Precedent can be found in the way in which the United States deals with monetary policy through the Federal Reserve Bank, its board of governors, and its chairman. With powers independent of both Congress and the White House, and with its attenuated terms of office, the FRB has consistently proven that it is not at the mercy of any branch of government but is dedicated to the long-term needs of the nation.

A similar institution could be erected for the country's industry—chartered to keep the United States a major presence in all key technologies and charged with sufficient powers to support the emergence of those technologies but not enough to interfere with them once under way. It must be impossible for such an agency to micromanage the new technologies; instead it must foster competition, entrepreneurship, and the other features of a free market. Otherwise, this agency will become yet another instrument for preserving the status quo, performing the latter-day equivalent of subsidizing the vacuum tube business long after the transistor has been invented.

In practice, this agency would select the industries important to the future of the country, then make investments and provide incentives to encourage their development. We already know what most of those technologies are; and one of the agency's tasks would be to identify others the instant they appear. We even have a prototype for just such an agency: DARPA (the Defense Advanced Research Purchasing Agency), which, for the last thirty years, has been chartered to identify key emerging military technologies and subsidize them. Despite being chronically underfunded and subject to all the failings of the Defense

Department, DARPA has managed over the years to play a key role in the creation of many of America's most successful industries, including commercial aircraft and semiconductors.

The proposed agency would be a commercial DARPA with more teeth. It would fund academic and industrial research and recommend tax and financial incentives for companies making investments in these target areas. It would encourage (through such funds as scholarships and endowments) training in disciplines that would support these industries so the necessary human resources would be available.

The Malcolm Baldrige Award is an excellent example of just such a government incentive program that is very inexpensive to run but probably has done more to improve the long-term competitiveness of our industry than any other government program. Here is a case of good industrial policy. This national recognition award, modeled after Japan's Deming Prize and designed to raise the quality of America's industrial output, deserves much of the credit for the progress made here in recent years.

In support of such a program, the federal government must also reexamine its antiquated antitrust laws, many of which are as outmoded as Fordism, Taylorism, and management hierarchies. In a world of international competition and free trade, it is difficult to envision many monopolies of the type that antitrust laws were designed to stop. Our companies need to pool resources in order to meet the threat of foreign competition.

An example of this type of cooperation is SEMATECH, a $200 million semiconductor industry consortium funded by the government and private industry aimed at improving the competitiveness of capital equipment suppliers to the semiconductor industry. SEMATECH required congressional approval—a convoluted and wasteful process— that would have been more quickly accomplished by a government agency (or simply a collection of corporations) operating under more realistic antitrust regulation. European nations have created a number of comparable industry consortia with far less difficulty and far more funding, including EUREKA ($5 billion annually), ESPRIT, JESSI ($4 billion), and RACE. Wrote the Lehigh University researchers in *21st Century Manufacturing Enterprise Strategy:*

> Creating standard cooperation models, certified in advance as legal, would go a long way toward making cooperation easier and more attrac-

tive. . . . More broadly, change is required in the prevailing attitude in American society towards anti-trust legislation, as well as in the legislation itself. The historical foundation of anti-trust legislation in the U.S. has been superseded by events.[7]

# Tax Policy

The tax code has determined much of the behavior of U.S. industry for the last century. It could be used much more effectively to meet the needs of virtual corporations. In fact, the tax code as it is currently structured does a great deal to undermine these needs.

The objectives of a good tax code should be to generate revenue for the government to carry out necessary programs, encourage socially desirable activities and discourage others, and to be fair in the manner in which it places a burden on individuals and institutions. Using the tax code as a way to directly redistribute wealth is not a legitimate objective. The only effective wealth redistribution system is prosperity—any other workable policy, as Drucker has noted, devolves to the use of inflation to expropriate the middle class, "destroying productivity" in the process.[8]

If one were to redesign the tax code today in light of the needs of virtual corporations, it would encourage long-term investment, savings, education, and research, and it would discourage consumption of non-basic and luxury items and precious resources such as oil and water. This could be accomplished so that the tax incentives were not regressive.

One of the primary goals of a new tax policy would be to encourage a long-term view toward investment and discourage investment speculation. A way to do this would be to steeply tax all short-term capital gains and reduce or eliminate the tax on long-term gains for all investors. Such a policy would encourage both managers and investors to adopt a more distant business horizon. It would as well discourage speculation by large tax-free institutions, such as pension funds. It seems likely that such a tax could be made revenue neutral from the increased taxes on speculative transactions by nonprofit institutions and from raising short-term taxes to offset the reductions elsewhere.

The government could choose as well to tax consumption. It could

255

place luxury taxes on all nonessential items costing more than a few hundred dollars. Those taxes could escalate with price. The government could also encourage savings by increasing the size of retirement savings exclusions and eliminating or reducing the tax on interest and dividend income. By coupling these programs with some form of income tax reduction for lower- and middle-income earners these taxes could be made nonregressive.

There are a number of products and technologies that are no longer produced in our country but are important to the creation of virtual products. The United States needs domestic access to many of these consumer electronic technologies, as they will be the basis for many of the critical computation products and mass-customized manufacturing systems of the future. At present there is no domestic manufacture of many of these products. Instead, domestic purchase of imported VCRs, camcorders, low-end fax machines, pocket organizers, cameras, and other worthy but unessential consumer products constitutes a major component of the nation's trade imbalance.

If the goal is to restore this technology (and the skills that emanate from it) to the United States, then foreign manufacturers must be impelled to move onshore. In other words, some form of domestic technology content regulation must be put in place. One way to do this would be to levy a sufficiently large excise tax on these items to greatly curtail their sale in the United States. This tax would then be lifted, for all competitors, when domestic design and manufacture of these products are established with sufficient local content. If this suggestion seems radical, keep in mind that it is precisely what happened, albeit unconsciously, when Japanese electronics and automobile companies feared the imposition of U.S. tariff barriers and quickly emplaced plants and factories throughout the United States. The result has been beneficial for both nations.

Sweeping tax code revisions are always unpredictable. Motivations change and in the process drive the economy in new, unexpected, and sometimes even harmful directions. Individuals and corporations will always figure out how to exploit the changes in ways never imagined by government. That is why an attitude of social *kaizen* is so important. In tax policy as elsewhere, we should pick a direction and move toward it systematically and in small steps, constantly evaluating what works and what does not. In this way, the government can learn as it goes and avoid the haphazard and contradictory patterns that have so often characterized our tax policies in the past.

# Education

In 1991 the National Association of Manufacturers, working with consultant Towers Perrin, surveyed four thousand NAM members about the quality of their workers' job skills. The results, publicized in early 1992 on the same day that the speaker of the Japanese house of representatives decried American workers as lazy, selfish, and illiterate, were singularly depressing:

> The survey revealed, for example, that the average manufacturer rejects five out of every six candidates for a job, and that two-thirds of companies regularly reject applicants as unfit for the work environment. A third of the companies said they regularly reject applicants because they cannot read or write adequately, and one-fourth reject them because of inabilities with communications and basic mathematics.
>
> As for those already employed: More than half the companies reported major employee skills deficiencies in basic math, reading and problem-solving. . . .
>
> Thirty percent of companies couldn't reorganize work activities because employees couldn't learn new jobs; 25 percent said they were unable to improve product quality because workers couldn't learn the needed skills.[9]

The virtual corporation will deeply depend upon a skilled and trained work force that is not only literate but capable of decision making and self-direction. Employees will have to participate in teams, analyze problems, and propose and implement solutions. This will never happen if, as the NAM report concluded, "60 percent of new jobs will require more than a high school education. However, 70 percent of new entrants into the workforce will have less than a high school degree."[10]

It is therefore crucial that government find ways to improve primary and secondary education in the United States. It is difficult to imagine the country doing this simply by returning to basics, raising teacher salaries, or cutting classroom sizes. All these programs would certainly help but are contingent upon resources that probably will not be found.

Education in the United States needs its own transformation, remarkably synchronous with the revolution required in American business. Education theorist Theodore Sizer has called for the U.S. educational system to abandon its century-old notions of "One Best Curriculum and One Best Pedagogy and One Pace of Learning and One Best Test"[11] and move to a more flexible, adaptive system built on trust in which "the decentralizing of substantial authority to the persons close to the students is essential."[12] Economist Robert Reich agrees, "We need to push the responsibility down to the front line, toward teachers and principals."[13] As in industry, the best hope of attacking the problem lies in applying the computer to education.

We have no illusion that computers are the panacea for poor education, but certainly they can be an invaluable supplement—especially now that the cost of computation and of multimedia systems is cheap enough to make widespread adoption practical. Computerized texts have been developed that have demonstrated effectiveness, especially those that adapt to the diverse needs of students. Some states today even permit textbook budgets to be spent on educational software. Given what are likely to be permanent constraints on education budgets, computers offer the one hope of freeing teachers from the daily burden of teaching basic skills and allow them to dedicate themselves to the more elusive (and rewarding) challenge of creating lifelong learners.

Just creating flexible, curious, and competent minds will be enough of a challenge for schools. Companies will need employees with job-specific skills. It is their obligation to provide much of this training, yet it will not happen as often as it is needed if companies view their work forces as transient. Why make a training investment in an employee who will use it to seek a better job elsewhere?

One solution to this is tax credits covering a substantial portion of employee training, say, 90 percent of training costs for the first year of employment and falling 10 percent per year thereafter. This would recognize the overall social benefit of this private training, while at the same time, by not offering complete reimbursement, provide some protection against abuse.

The government should as well actively ensure that adequate numbers of trained college graduates are available to our society. All college diplomas are not equal. We currently have too many doctors and lawyers and too few teachers and scientists. One way to shape the composition of graduating classes to more closely meet society's needs is to offer targeted incentives; for example, by making loans and scholarships more readily available to students pursuing certain degrees. As the needs of society change, the bias in such a program could be altered.

There is considerable precedent for this. ROTC programs have long been used to produce trained military officers. If indeed the war of the twenty-first century will be for economic survival, then our government should be willing to commit the resources to produce future leaders in that combat.

The higher education system in the United States may be the envy of the world, but it is not doing the job required to train students for work in virtual corporations. Most colleges and universities in this country reinforce the old notion of the two cultures, science and the humanities. Liberal arts programs produce students who are technically illiterate, incapable of dealing with the technical content of a modern economy. Meanwhile, science and engineering majors, by being spared a leavening of humanities, are woefully unprepared for the social structures and interpersonal relations that are at the heart of virtual corporations. Says Reich:

> Technological literacy is fundamental. The emerging global economy requires people at all levels who understand technology, design engineering and manufacturing engineering, energy, production, and so on. . . . If you don't have the skills, you are in competition with millions of others worldwide eager to work for less. Blue collar jobs at $10 and $20 an hour are vanishing from these shores. Our unskilled workers are in competition with unskilled workers everywhere.
>
> More and more of the jobs available to unskilled American workers are confined to the local service economy, where it's difficult to make much money. Many Americans still assume that we're living in the 1950s, when a high school graduate could still get a good job in the factory in the next town. That era just doesn't exist any more.[14]

Even our graduate schools of business, for all their fame, are doing little to prepare the future leaders of American industry to run virtual corporations. On the contrary, at a time when multidisciplinary competence, technical acumen, and leadership skills are needed, this country's B schools continue to instruct future executives in traditional management styles and career specialties and, at best, give only a passing nod to the vital role of information.

The greatest danger in not properly training our young people to participate in a twenty-first-century economy is that of creating a permanent army of disaffected unemployed, perpetually frozen out of the progress occurring around them. Not only would such a group be unhealthy to the society as a whole (as we can already see with our current underclass), but it would, if it comprised a sufficiently large percentage of the population, ultimately negate whatever gains were made by the new business revolution.

A literate person in the year 2000 will have to be both technically and socially competent. That means engineers must have an understanding of cultural anthropology and an appreciation of literature, and that English majors need to be skilled with the personal computer and understand science and technology. This is a tall order, but in some respects it harkens back to the college education of a century ago, when generalism was more valued than specialization. In the virtual corporation, to have just one specialty is to be a burden on the rest of the organization.

# Infrastructure

One of the biggest tasks facing the United States in the years to come will be the reconstruction of its deteriorating infrastructure. From the end of the Second World War through the 1960s the United States spent 4 percent of its GNP on infrastructure. Since then it has spent just 1 percent, and a reckoning is approaching.[15]

The intuitive answer to this problem is to spend the huge sums needed to rebuild the nation's roads and highways. But this is not necessarily the correct solution. If traditional corporations were fueled by oil—for trucking, power, and employee commutes—virtual corporations will be driven by massive amounts of increasingly sophisticated information. Thanks to government research, corporate investment, and private initiative (such as Internet), the United States is the world's leader in high-speed computer networking. It is a lead we dare not lose.[16]

In a twist of history, it has been estimated that the cost of either rebuilding the nation's highways or installing a national fiber-optic network will be about $200 billion. Thus, we stand at a crossroads. It would be a dangerous mistake, we believe, when the nation is faced with either rebuilding its roads or installing a high-speed communications grid, to give priority to the former. That, once again, would be merely improving the past, not embracing the future.

Here again is where intelligent industrial policy can be useful. As late as 1987, AT&T predicted it would take until 2010 to convert its entire long-distance network to digital switching. Thanks to competition from the likes of U.S. Sprint, the job was done by 1990.[17] Private competition, spurred by government incentives and investment, might put the data grid in place in a fraction of the expected time with much less burden on the taxpayer.

Perhaps also making such a choice more palatable is the possibility, as George Gilder has suggested, that the advent of high-quality multimedia data communications may obviate the need for a sizable fraction of current commuting and business travel time. The government might also provide incentives for the privatization and expansion of mass transit programs. All of these efforts could buy the nation some time, enabling it to gather the funds needed for bridge and highway improvement. Nevertheless, good roads or bad, whenever the communications grid needs upgrading (such as to the Broadband Integrated Service Digital Network early in the next century), it must take precedence. That will be the choice our best competitors will make. Already Nippon Telegraph and Telephone has announced it will wire Japan for fiber optics by 2015.[18]

# Legal Reform

The current legal system is no longer protecting the rights of consumers but draining their blood. It is estimated that the current tort system costs the country $300 billion a year in direct and indirect costs. This amount is equivalent to the total defense budget. Anecdotal evidence is even more alarming. For example, malpractice insurance adds $300 to the cost of a birth in New York City.[19] It has been estimated by the Rand Corporation that $2 are spent on the legal system for every $1 delivered to an injured party. And, thanks to legal judgments, the U.S. small plane industry has essentially disappeared—ironically leaving enthusiasts to fly older, less safe planes.

Three hundred billion dollars is 6 percent of the nation's gross national product—enough to make *Forbes* complain: "Roll over, Wall Street. Meet the real champions of the great American greed game: plaintiff attorneys—lawyers who specialize in suing." These plaintiff attorneys alone have a combined estimated annual income from contingency fees of more than $10 billion.[20] Certainly most of that money would be better spent on education, research, or improving the infrastructure.

These costs could be dramatically reduced by forcing losers in tort cases to pay some type of fine for the inconvenience they have caused. One notion has been to have the losing party assume all of the other side's costs. But there is a validity to the argument that this would have a chilling effect upon individual litigants bringing suit against giant corporations. A more practical example can be found in the medical insurance industry, where the new practice of having company employees pay a small fraction of their medical costs has sizably reduced benefits filings. By the same token, requiring the loser in a tort case to pay, say, 10 percent of his or her attorney's costs might well knock off a majority of nuisance suits and settlement-fishing expeditions, while still preserving meaningful litigation undertaken by serious plaintiffs.

Proposals have been made for alternative forms of dispute resolution to cut the cost of delivering compensation to the injured party. One idea with considerable appeal is that punitive damages should be limited to some reasonable multiple of actual damages. Another involves taking from juries the responsibility of setting damages and giving it back to judges. Both have appeal—after all, courts are meant to render fair judgments, not serve as instruments of personal vengeance. Such legal constraint becomes especially important in a society that is built upon interdependence and mutual trust.

# Labor

Labor organizations based on the assumption that management and labor are blood adversaries are as obsolete as managements that believe they can continue to exploit workers. Both will disappear because the institutions in which they operate will be the business failures of the twenty-first century.

The role of unions and worker organizations of the future is to improve job skills, to help management improve efficiency, and to ensure that workers are treated fairly and share in the success of their employers. The issue is not whether unions are strong or weak but whether labor and management are capable of teamwork. If they are not, both will fail.

From management's point of view, it should either strive to form a team with labor or it should attempt to rid itself of unions. There can be no middle ground. While this position may appear to be anti-organized labor, we hasten to add that the converse is also true: if the union is operating in good faith and its efforts are impeded by management, the company's board of directors should fire the executives.

As was discussed in chapter 9, there is some evidence that the most powerful combination for creating a virtual corporation is one featuring both an enlightened management and a union. This suggests that in most cases where unions already exist, the optimum approach is to find ways to cooperate, such as through shared worker training.

Unfortunately, in the rapidly paced world of the virtual revolution, a recalcitrant and reactionary union cannot be suffered for very long. Ridding a company of a union is a divisive and destructive process. But living with one that is anti-company is even more disastrous. Such a union will soon kill the company and itself—leaving the poor workers as the ultimate victims.

# Trust and Teamwork

One of the recurring themes throughout this book has been the need for trust and teamwork among government, business, and workers. There is, of course, no law that can be passed that will force this to happen. But it is possible to create an environment in which such relationships are likely to occur.

The move toward a more coherent industrial policy will begin the process of bringing government and industry together. Diminishing the pernicious distortions of the legal system will force individuals to solve their problems in nonlitigious ways. Helping managements to rid themselves of uncooperative labor groups, and vice versa, will mean those that remain will work as teams. Providing tax incentives that orient management toward long-term goals will improve relationships with customers, employees, and suppliers. Encouraging business to invest heavily in training its workers will make it more dependent on them and more interested in developing employee trust.

In the future, treating customers, suppliers, and employees fairly will be vital to the success of a business. Companies such as General Motors are today paying the price for years of abuse of these relationships. In the future, competition will be so fierce and the need for trust and teamwork within the virtual corporation so vital that the market will exact its retribution with shocking speed. In this environment, under the threat of annihilation, we will have to cling to one another for safety.

Ben Franklin's words at the birth of this country seem to have a special currency: "We must all hang together or assuredly we shall all hang separately."

## The Next Industrial Era

For the United States there is no other choice but to lead the way into the new era. Our nation could not long stand the diminishment and the economic colonialism that would come from being an also-ran in the world's economy.

On the positive side, the revolution driven by virtual products represents a clean slate, a way to break with mistakes of the past and begin anew. Says the Lehigh University report: "The fact that all of the world's leading manufacturers have to build a new infrastructure to make the transition from mass production to agile manufacturing provides a unique opportunity for U.S. industry to regain the leadership it lost in the 1970s and '80s."[21]

Happily, despite our many problems, the United States is in many ways the best suited for the new business revolution. Our technological innovation remains the best on the globe. Our productivity is also the world's highest, though no longer the fastest growing. Finally, we remain the world's greatest economic, military, and political power.

There are other, more subtle, advantages as well. We are a multi-ethnic society that has embraced the most adventurous and creative men and women of the world. Our university system, especially its graduate programs, are the envy of other nations. And, certainly, the history of this nation has proven the extraordinary resilience and strength of our political system.

Ultimately, however, America's greatest advantage is its storied individualism, and the entrepreneurship that arises from it. It was this entrepreneurship that put the United States at the forefront of other industrial transformations of the past two centuries and, if properly supported, promises to do so again. By comparison, the Japanese have few of these prerequisites. Says the Lehigh report:

> The new system will depend on spontaneous work force initiative and on innovative creativity at the operational level of an enterprise. Neither of these have been noteworthy characteristics of Japanese industrialization, or society. The system will depend on distributed authority and globally decentralized manufacturing facilities, but the Japanese have had difficulty treating other nationals as peers. Both at home and abroad, Japanese managers will have to deal with women who will merit, and expect, promotion to positions of authority, yet another problem of Japanese culture. Finally, the successful implementation of the new technologies will be critically dependent on the software linking human workers at all levels with a national information network capable of controlling manufacturing operations. Software development has not been a strength of the Japanese computer industry to date.[22]

As for the United States, our greatest handicap may be the legacy of past success. We have been so wealthy for so long that we have not had to make the tough choices. For four decades now, we have believed we could fight wars and live comfortable lives at home. Social programs have promised our citizens incomes whether or not they made an effort to contribute to society. We guaranteed comfortable retirements to our citizens whether they saved or not. We promoted and graduated students rather than dealing with the truth of their failure.

It is now clear that to capture the future we will have to work as hard as we have at any time in our history. We are competing for our existence against others who want to live better than they do today and who are willing to work very hard to reach that goal.

Because of the diffusion of technology around the world, other nations can now build many of the same products that we can build. This is no longer the world of the 1950s where only a few advanced countries could make steel, cars, electronics, airplanes, computers, copying machines, and machine tools. If we wish to prosper we have to find ways to build products that offer more value. The virtual corporation offers us that opportunity.

In the midst of turmoil, it is easy to become overwhelmed by our problems. They are very large, but they are not unsolvable. Imagine how the world must have looked to a Japanese or German citizen in 1945. Their economies had been devastated by bombs as well as arrogance. Ours is a victim of mere complacency.

What is now required is that we commit ourselves to winning the race, one in which there are no leaders today and which two economies are best positioned to enter: ours and Japan's. And, as if preordained for a classic conflict, each of these two enjoys vital abilities the other lacks.

The United States has the opportunity to race ahead and lead the world into the future created by virtual corporations. But for an opportunity to become an advantage requires will. It remains to be seen if we Americans are willing to overcome the divisive forces in our society, including race, ethnicity, and class, for the sake of our nation's future. Do we still have our historic courage to embrace change?

# The Last Requirement

A final question must be asked. In the end, will an economy based upon virtual products result in a better society? Ultimately, after all the losses and gains, will it be for the good? *Is the virtual corporation virtuous?*

This is not an idle question. The monetary expenditure and human energy required of a society to transform its economy will be so great that a morally ambiguous goal will not be worth the effort.

There is as well the issue of who will work. Factories in virtual corporations will be so efficient that not as many individuals will be needed to produce the goods our society needs. Just as there are no longer many people required to work on the farm, so will the need for numerous factory workers vanish. The idea of sharing the work by moving to shorter work weeks is not a practical one either. The worker in the virtual corporation will require so much skill and training that it will not be realistic to educate numerous people to fill the same job.

Therefore, it is likely that in the future perhaps 85 percent to 90 percent of us will work in service jobs, many of them within manufacturing companies. But there will be a difference between the economy of tomorrow and the service economy of today. Tomorrow's will be an economy with a strong agricultural and manufacturing base capable of sustaining a service infrastructure. It will no longer be a hollowed-out economy but a complex, rich one.

Certainly it can be argued that all business and social revolutions ultimately are an improvement over what came before. That's why they occur. Too often books about business assume this answer and go on with their celebrations of the halcyon days to come. The real truth is that all change comes with great cost, for the guilty and sometimes for the innocent. The approaching business revolution will be no different.

But any system of human organization that values both the freedom and the power of workers to control their lives, demands an educated population, answers the unique needs of individuals, is built on trust and cooperation rather than distrust and confrontation, and rewards long-term dedication rather than short-term manipulation, surely must be considered an improvement to the common good.

Furthermore, the evolution from virtual product to virtual corporation to a new economy does, by the evidence to date, represent a virtuous cycle—that is, a positive upward spiral. As virtual corporations become more pervasive many of the current transitional problems and distortions that currently affect our society will be erased.

But none of these salutary results are automatic. For the virtual corporation to appear here, now, will require all of its employees to revise the way in which they deal with one another and with the outside world. To live in the new world created by virtual products, all of us must make changes in our lives.

267

In 1848, Karl Marx predicted that the business and social revolution occurring around him would lead to a revolt by the workers against their capitalist masters. That revolt never occurred. Marx was wrong because he failed to predict that the mass production revolution would also raise the quality of life for the workers. He was also wrong because the societies themselves in time recognized the inequities of the new system and set about rectifying them through charitable institutions, vocational schools, and public health and welfare.

Like the industrial age before it, the emerging era has the potential to raise the quality of life for everyone to unprecedented levels. Like earlier social transformations, it will likely reward a whole new class of individuals. But, also like previous economic transformations, it will leave some behind—people who cannot cope with the new responsibilities, the rapid pace of change, and the demands for mental adaptability. In the frantic pace of life in this new economy, in the full-time job of maintaining established relationships, it will be easy to forget these others.

A just society, a virtuous society, will tend to the needs of the disenfranchised. Thus, the last requirement of the coming business revolution is that it also exhibit the quality of mercy.

# Notes

## Chapter 1: A New Kind of Business

1. James Burke, *The Day the Universe Changed* (Boston: Little, Brown & Co., 1985), p. 190. This source was used for both the quote and the story of the steam engine.

2. D. C. Gazis, "Brief Time, Long March," in *Technology 2001* (Boston: MIT Press, 1991), p. 44.

3. Louis Trager, "Taco Bell Express Offers Fast Food at a Frantic Pace," *San Francisco Examiner,* 10 November 1991, p. E-1.

4. We thank Earl Hall for this elaboration of the definition.

5. Stan Davis and Bill Davidson, *2020 Vision* (New York: Simon & Schuster, 1991), p. 58.

6. We note that the only other published appearance of the phrase "virtual company" that we have found is in *21st Century Manufacturing Enterprise Strategy,* published by the Iacocca Institute of Lehigh University in November 1991. The authors, Roger Nagel and Rick Dove, apparently came to the phrase (and a number of other similar notions about the future of manufacturing) independently and their use is somewhat different than ours. For Nagel and Dove, a virtual company is created by "selecting organizational resources from different companies and then synthesizing them into a single, electronic, business entity." While this is an appealing idea, we believe it will be an uncommon business structure for many years to come.

7. Alvin Toffler, *The Third Wave* (New York: William Morrow, 1980), pp. 271–273, 282–305.

8. Advertisement for HP model 700, in *Electronic Engineering Times,* 13 January 1992, p. 29.

9. Earl Hall, personal interview.

10. Alfred D. Chandler, Jr., *The Visible Hand* (Boston: Belknap/Harvard University Press, 1977), section 2, pp. 79–206.

11. See Robert Reich, *The Work of Nations* (New York: Knopf, 1991), pp. 81–86. See also George Gilder, "The Message from the Microcosm," chap. 1 in *Microcosm* (New York: Touchstone, 1989).

12. Stephen S. Cohen and John Zysman, *Manufacturing Matters* (New York: Basic Books, 1987), pp. 12–13.

13. Chandler, *The Visible Hand*, p. 110.

14. For example, Professor Gary Becker, as quoted in *Business Week*, 27 January 1986, p. 12.

15. Michael L. Dertouzos, Richard K. Lester, and Robert M. Solow, *Made in America* (Cambridge: MIT Press, 1989), pp. 12–14.

16. Paul Krugman, *The Age of Diminished Expectations* (Cambridge: MIT Press, 1990), p. 3.

17. Dertouzos, Lester, and Solow, *Made in America*, p. 29.

18. Krugman, *Diminished Expectations*, p. xi.

19. Bruce Steinberg, "Raising Productivity Will Be Tough in '90s," *Wall Street Journal*, Outlook section, 10 December 1990, p. 1A.

20. "BRIE Globalization and Production," conference report, 15–16 April 1991.

## Chapter 2: An Emerging Idea

1. Glenn Haney, personal interview.

2. Stan Davis and Bill Davidson, *2020 Vision* (New York: Simon & Schuster, 1991), p. 58.

3. Wilf Corrigan, personal interview, March 1991.

4. Chandler, *The Visible Hand*, p. 280.

5. Joseph D. Blackburn, ed., *Time-Based Competition* (Homewood, Ill.: Business One/Irwin, 1991), p. 201.

6. James P. Womack, Daniel T. Jones, and Daniel Roos, *The Machine That Changed the World* (New York: Rawson Associates/MacMillan, 1990), p. 111.

7. Ibid., p. 124.

8. Gail E. Schares and John Templeman, "Think Small," *Business Week*, 4 November 1991, p. 62.

9. George Stalk and Thomas Hout, *Competing Against Time* (New York: Free Press, 1990), pp. 2–37.

10. Ibid., p. ix.

11. Roger Nagel and Rick Dove, *21st Century Manufacturing Enterprise Strategy* (Lehigh, Pa.: Iacocca Institute of Lehigh University, 1991), p. 36.

12. Ibid., p. 4.

13. Ibid., pp. 35–36.

14. Interview with Jim Morgan, 3 February 1992.

15. Ramchandran Jaikumar, "From Filing and Fitting to Flexible Manufacturing: A Study in the Evolution of Process Control," paper written Feb. 1988, presented at the Conference of Global Changes in Production and Manufacturing, U.C. Berkeley April 14–16, 1991.

16. Jim Carmichel, "New Guns the Way They Used to Be," *Outdoor Life*, May 1988, p. 72.

17. Ibid., p. 73.

18. Burke, *Day the Universe Changed*, pp. 111–112.

19. Daniel Melcher and Nancy Larrick, *Printing and Production Handbook*, 3rd edition (New York: McGraw-Hill, 1966), pp. 246-248.

20. Dirk Hanson, *The New Alchemists* (Boston: Little, Brown & Co., 1982), pp. 59–60.

21. Source for figures: U.S. Department of Transportation. Most of the information in this section is based upon the authors' own experience working in or reporting on the electronics industry.

22. Louis Uchitelle, "Airlines Off Course," *New York Times Magazine.* Reprinted in the *San Francisco Chronicle,* 15 September 1991, p. 7.

23. Unless otherwise noted, the history of the SABRE system comes from James L. McKenney, Duncan G. Copeland, and Richard O. Mason, "American Airlines SABRE System," HBR Working Paper 92–106 from draft presented to the History of Computing Conference on 24 June 1991.

24. "American Airlines SABRE (A)," Harvard Business School Case No. EA-C 758, 1967, p. 6. Quoted from "American Airlines SABRE System."

25. "The Cautious Pioneer," *Forbes,* 1 June 1956, pp. 3–5. Quoted from McKenney et al., "American Airlines SABRE System."

26. The Charles Ammann story can be found in his "Airline Automation: A Major Step," *Computers and Automation,* August 1957, p. 2.

27. W. R. Plugge and M. N. Perry, "American Airlines 'SABRE' Electronics Reservations System," AFIPS Conference Proceedings, Western Joint Computer Conference, 1961, p. 594. Quoted from McKenney et al., "American Airlines SABRE System."

28. R. F. Burkhardt, "The SABRE System: A Presentation," 1 October 1964, pp. 2–3. Quoted from McKenney et al., "American Airlines SABRE System."

29. Jim Bartimo, "Wanted: Co-Pilots for Reservation Systems," *Business Week,* April 9, 1990, p. 79.

30. Uchitelle, "Airlines Off Course," p. 10.

31. Kenneth Labich, "How Airlines Will Look in the 1990s," *Fortune,* 1 January 1990, p. 51. This number compares with 4.6 percent for all U.S. manufacturers.

## Chapter 3: Powers of Information

1. Womack, Jones, and Roos, *Machine That Changed the World,* pp. 67, 182.

2. Ibid., p. 189.

3. Taiichi Ohno, *Toyota Production System* (Cambridge, Mass.: Productivity Press, 1988).

4. Maryann Keller, *Rude Awakening* (New York: Harper Perennial, 1989), p. 127.

5. Womack, Jones, and Roos, *Machine That Changed the World,* p. 173.

6. Ibid., p. 118.

7. Joseph D. Blackburn, "The Quick Response Movement in the Apparel Industry: A Case Study in Time-Compressing Supply Chains," chap. 11 in *Time-Based Competition* (Homewood, Ill.: Business One/Irwin, 1991).

8. "Best Practice Companies," *FW,* 17 September 1991, p. 36.

9. Ibid.

10. Blackburn, "The Quick Response Movement," pp. 254–255.

11. Ibid.

12. Ibid., p. 258.

13. Chandler, *The Visible Hand,* p. 280.

14. Ibid., p. 94.

15. "GM Will Run Big Car Group Like Japanese," *Wall Street Journal,* 10 June 1991, p. A3.

16. Rosabeth Moss Kanter, "Attack on Pay," *Harvard Business Review,* Mar.–Apr. 1987, p. 60.

17. Brian Dumaine, "The Bureaucracy Busters," *Fortune,* 17 June 1991, p. 46.

18. John S. McLenahen, "Factory Automation," *Industry Week,* 20 June 1988, p. 44.

19. Gary H. Anthes, "U.S. Sees Industry Growth in '92," *Computerworld*, 6 January 1992, p. 88.

20. Standard & Poor's, *Industry Surveys*, 17 October 1991, p. C96.

21. Womack, Jones, and Roos, *Machine That Changed the World*, p. 13.

22. Ohno, *Toyota Production System*, from the Publisher Forward by Norman Bodek, p. ix.

23. Quoted by Gary Loveman, "Why Personal Computers Have Not Improved Productivity," minutes of Stewart Alsop 1991 computer conference, p. 39.

24. Ibid.

25. Author interview with Michael Borrus.

26. Paul A. David, "Computer and Dynamo: The Modern Productivity Paradox in a Not-Too-Distant Mirror," CEPR pub. no. 172, July 1989, from the abstract.

27. Ibid., pp. 16, 20.

28. Chandler, *The Visible Hand*, section 2.

29. Per Barry D. Olafson, Molecular Simulations.

30. George Tibbits, "Change Is in the Air as Boeing Builds New 777," *San Francisco Examiner*, 6 October 1991, p. E3; and "Picture Perfect," photo essay, *Business Week/Quality 1991*, p. 96.

31. See the Tacoma Narrows Bridge story in chapter 5.

32. Kathleen Bernard, "Ordering Chaos," in *Technology 2001* (Cambridge, Mass.: MIT Press, 1991), p. 82.

33. This appears to be what is going on at the very secretive Japanese robotics firm Fanuc.

34. Thomas M. Rohan, "Factories of the Future," special section, *Industry Week*, 21 March 1988, p. 44.

## Chapter 4: The Upward Curve of Technology

1. Robert Reich, *The Next American Frontier* (New York: Times Books, 1983).

2. Ruth D'Arcy Thompson, *D'Arcy Wentworth Thompson: The Scholar-Naturalist, 1860-1948* (New York: Oxford University Press, 1958). Quoted by Loren Eiseley in *The Night Country* (New York: Charles Scribners Sons, 1971), p. 135.

3. Gino Galuppini, *Warships of the World* (New York: Military Press, 1989), p. 16.

4. "Shipping," in *The American Peoples Encyclopedia*, vol. 17 (Chicago: Spencer Press, 1953) p. 599.

5. Hendrik Willem van Loon, *The Story of Mankind* (New York: Liveright, 1984), p. 68.

6. Galuppini, *Warships of the World*, p. 56.

7. Chandler, *The Visible Hand*, section 2, pp. 79–206.

8. Peter Wyden, *Day One: Before Hiroshima and After* (New York: Simon & Schuster, 1984), p. 213; and Richard Rhodes, *The Making of the Atom Bomb* (New York: Simon & Schuster, 1986), p. 711.

9. Author interview with Dr. Gordon Moore for Silicon Valley Report television program.

10. Ibid.

11. Gazis, "Brief Time, Long March," p. 48.

12. Linda Runyan, "40 Years on the Frontier," *Datamation*, 15 March 1991, p. 34.

13. Ibid., p. 36.

14. MIPS Computer Systems prospectus, 7 November 1989, p. 23.

15. Terry Costlow, "The Drive for Capacity," *Electronic Engineering Times*, 14 October 1991, p. 1.

16. Hanson, *The New Alchemists*, p. 60.

··············································

17. B. K. Ridley, *Time, Space and Things* (New York: Penguin Books, 1976), p. 52.

18. D. Bruce Merrifield, "Management of Critical Technologies," Wharton School of Business, 1991, p. 1.

19. Ibid., Appendix A. Here is the complete list:
    Materials
        Materials synthesis and processing
        Electronic and photonic materials
        Ceramics
        Composites
        High-performance metals and alloys
    Manufacturing
        Flexible computer-integrated manufacturing
        Intelligent processing equipment
        Micro- and nanofabrication
        Systems management technologies
    Information and Communications
        Software
        Microelectronics and optoelectronics
        High-performance computing and networking
        High-definition imaging and displays
        Sensors and signal processing
        Data storage and peripherals
        Computer simulation and modeling
    Biotechnology and Life Sciences
        Applied molecular biology
        Medical technology
    Aeronautics and Surface Transportation
        Aeronautics
        Surface transportation technologies
    Energy and Environment
        Energy technologies
        Pollution minimization, remediation, and waste management

20. Jerry Sanders, quoted from the Ben Rosen Technology Conference, New Orleans, June 1980.

21. L. Saunders, "Why April 15 is Getting Worse and Worse," *Forbes*, 18 March 1991, pp. 84–85.

22. Robert R. Gaskin, "Paper, Magnets and Light," *Byte*, November 1989, pp. 391–392.

23. The product was introduced in early 1992 by Maxtor.

24. Gaskin, "Paper, Magnets and Light," p. 399.

25. Esther Dyson, "Computers Programming Computers," *Forbes*, March 6, 1989, p. 137.

26. $100 billion estimate per Peter Coy, "Why the Highway Won't Reach Home Just Yet," *Business Week*, 16 September 1991, p. 112.

27. W. R. Johnson, Jr., "Anything, Anytime, Anywhere," in *Technology 2001* (Cambridge, Mass.: MIT Press, 1991), p. 153.

28. Keller, *Rude Awakening*, p. 204.

## Chapter 5: The Future by Design

1. Womack, Jones, and Roos, *Machine That Changed the World*, pp. 21–25.

2. Howard Rheingold, *Virtual Reality* (New York: Summit Books, 1991), p. 30.

3. Ibid.

4. From a speech by Robert Reich.

5. Rick Whiting "Product Development as a Process," *Electronic Business,* 17 June 1991, p. 31.

6. Ibid., p. 29.

7. Stalk and Hout, *Competing Against Time,* p. 211.

8. William H. Davidow and Bro Uttal, *Total Customer Service* (New York: Harper & Row, 1989), p. 145.

9. Robert W. Burnett, "Concurrent Engineering," *IEEE Spectrum,* July 1991, p. 33.

10. Sharon Machlis, "Management Changes Key to Concurrent Engineering," *Design News,* 17 September 1990, pp. 36–37.

11. Whiting, "Product Development as a Process," p. 31.

12. Burnett, "Concurrent Engineering," pp. 24–25.

13. Womack, Jones, and Roos, *Machine That Changed the World,* pp. 116–117.

14. David Woodruff, "The Racy Viper Is Already a Winner for Chrysler," *Business Week,* 4 November 1991, pp. 36–38.

15. David Woodruff, "GM: All Charged Up Over the Electric Car," *Business Week,* 21 October 1991, p. 106.

16. Richard W. Griffith, director of Ford Microelectronics, "Perspectives on Innovation in Japanese Companies," DARPA/Sematech paper, July 1991, p. 40.

17. Rick Whiting, "Drawing the Road Map for Product Development," *Electronic Business,* 17 June 1991, p. 61.

18. James L. Nevins and Daniel E. Whitney, *Concurrent Design of Products and Processes* (Cambridge, Mass.: McGraw-Hill, 1989), pp. 53—54. It is interesting that within two years so many other German car makers had adopted the cone-point screws that their price fell to that of the flat-tips.

19. John Teresko, "A Report Card on CAD/CAM," *Industry Week,* 19 March 1990, p. 50.

20. Ibid.

21. Machlis, "Key to Concurrent Engineering," p. 37. The storage device is the RA90.

22. Hugh Willett, "What Drives Quality at Intel?" *Electronic Business,* 7 October 1991, pp. 29, 36.

23. Nevins and Whitney, *Concurrent Design,* pp. 48–52.

24. CIMLINC press release, 12 November 1991.

25. Lewis H. Young, "Product Development in Japan: Evolution vs. Revolution," *Electronic Business,* 17 June 1991, p. 75.

26. Stalk and Hout, *Competing Against Time,* pp. 111–114.

27. Ibid., p. 114.

28. Teresko, "A Report Card on CAD/CAM," p. 44.

29. Ibid.

30. Ibid., p. 45.

31. Ibid.

32. George Gilder, *Microcosm* (New York: Simon & Schuster, 1989), pp. 199–203.

33. From interviews at LSI Logic, April 1991.

34. Teresko, "A Report Card on CAD/CAM," p. 44.

35. Rheingold, *Virtual Reality,* pp. 26–27.

36. Otis Port, "Questing for the Best," *Business Week/Quality* 1991, p. 16.

37. Peter Drucker, "Japan: New Strategies for a New Reality," *Wall Street Journal,* 2 October 1991, p. A12.

38. R. S. Engelmore and J. M. Tenenbaum, "Toward the Engineer's Associate: A

National Computational Infrastructure for Engineering," White Paper, September 1990, p. 3.

    39. Ibid., p. 4. Referenced from E. A. Feigenbaum, P. McCorduck, and H. P. Nii, *The Rise of the Expert Company* (New York: Times Books, 1988).

## Chapter 6: The Machinery of Change

    1. Port, "Questing for the Best," p. 10.

    2. Earl Hall, "Competitive Manufacturing Infrastructure Issues for the 21st Century," prepared for the National Center for Manufacturing Sciences, March 1991, p. 3.

    3. Womack, Jones, and Roos, *Machine That Changed the World,* pp. 71–103.

    4. Stalk and Hout, *Competing Against Time,* p. 1.

    5. David Woodruff, "A Dozen Motor Factories—Under One Roof," *Business Week,* 20 November 1989, p. 90.

    6. Rohan, "Factories of the Future," p. 44.

    7. Cohen and Zysman, *Manufacturing Matters,* p. 131.

    8. Stanley M. Davis, *Future Perfect* (Reading, Mass.: Addison Wesley, 1987), p. 14.

    9. Chandler, *The Visible Hand,* pp. 236, 244.

    10. Womack, Jones, and Roos, *Machine That Changed the World,* p. 120.

    11. Technologic Partners Computer Letter, 8 July 1991, p. 5.

    12. Stalk and Hout, *Competing Against Time,* p. 1.

    13. Philip R. Thomas, *Competitiveness Through Total Cycle Time* (Cambridge, Mass.: McGraw-Hill, 1990), p. xii.

    14. Timothy D. Schellhardt and Carol Hymowitz, "U.S. Manufacturers Gird for Competition," *Wall Street Journal,* 2 May 1989.

    15. Ibid.

    16. Blackburn, *Time-Based Competition,* pp. 15–16.

    17. Rohan, "Factories of the Future," p. 42.

    18. Ibid., p. 42.

    19. Port, "Questing for the Best," p. 14.

    20. Jagannath Dubashi, Motorola Bandit story from "The Bandit Standoff," *FW,* 17 September 1991, pp. 48–50.

    21. "About Time," *The Economist,* 11 August 1990, p. 72.

    22. Stalk and Hout, *Competing Against Time,* pp. 35–36.

    23. Womack, Jones, and Roos, *Machine That Changed the World,* p. 13.

    24. Author conversation with Professor Benjamin Coriat.

    25. Womack, Jones, and Roos, *Machine That Changed the World,* p. 13.

    26. Ohno, *Toyota Production System,* p. xiii.

    27. Keller, *Rude Awakening,* p. 131.

    28. Ibid., pp. 126–128.

    29. Ohno, *Toyota Production System,* p. 22

    30. Ibid., p. 39.

    31. Ohno, *Toyota Production System,* pp. 25–26.

    32. Richard J. Schonberger, *Japanese Manufacturing Techniques* (New York: Free Press, 1982), p. 16.

    33. Yasuhiro Monden, *Toyota Production System* (Cambridge, Mass.: Industrial Engineering and Management Press, 1983), p. 4.

    34. *Purchasing,* 12 September 1985, pp. 21–23, and 15 January 1987, pp. 33–34; and Charles O'Neal and Kate Bertrand, *Developing a Winning JIT Marketing Strategy* (Englewood Cliffs, N.J.: Prentice-Hall, 1991), p. 11.

    35. Schonberger, *Japanese Manufacturing Techniques,* p. 12.

    36. O'Neal and Bertrand, *Developing a Winning JIT Marketing Strategy,* p. 13.

37. Interview with Tom Kamo.

38. Hall, "Competitive Manufacturing Infrastructure Issues," p. 17.

39. Ohno, *Toyota Production System,* p. 3.

40. J. M. Juran, *Juran on Planning for Quality* (New York: Free Press, 1988), pp. 4–5.

41. Schonberger, *Japanese Manufacturing Techniques,* pp. 71–73.

42. Philip Crosby, *Quality Is Free* (Cambridge, Mass.: McGraw-Hill, 1979), p. 38.

43. Philip B. Crosby, *Quality without Tears* (Cambridge, Mass.: McGraw-Hill, 1984), Chapter 1.

44. Ibid., pp. 39–41.

45. Both stories from Thane Peterson, "Top Products for Less than Top Dollar," *Business Week/Quality* 1991.

46. Jonathan B. Levine, "It's an Old World in More Ways than One," *Business Week/Quality* 1991, p. 27.

47. Ibid.

48. Joseph B. White, "Japanese Auto Makers Help U.S. Suppliers Become More Efficient," *Wall Street Journal,* p. A1.

49. Keller, *Rude Awakening,* pp. 206–208.

50. Ibid., p. 209.

51. Masaaki Imai, *Kaizen* (Cambridge, Mass.: McGraw-Hill, 1986), p. xxix.

52. Cohen and Zysman, *Manufacturing Matters,* p. 133.

53. Robert Neff, "No. 1—And Trying Harder," *Business Week/Quality* 1991, p. 20.

54. Cohen and Zysman, *Manufacturing Matters,* p. 132.

55. Ohno, *Toyota Production System,* p. 17.

56. Imai, *Kaizen,* pp. 30–32.

57. Ibid., p. 1.

58. Reza A. Maleki, *Flexible Manufacturing Systems* (New York: Prentice-Hall, 1991), p. 8.

59. Ibid., p. 14.

60. Ibid., p. 8.

61. Imai, *Kaizen,* pp. 25–27.

62. Ramchandran Jaikumar, "Postindustrial Manufacturing," *Harvard Business Review,* November-December 1986, pp. 69–76.

63. Donald Gerwin, "Manufacturing Flexibility in the CAM Era," *Business Horizons,* January-February 1989.

64. Maleki, *Flexible Manufacturing Systems,* pp. 10–11.

65. Brian M. Cook, "Flexible Manufacturing: Something That People Do," *Industry Week,* 5 November 1990.

66. Ibid., p. 43.

67. Hall, "Competitive Manufacturing Infrastructure Issues," p. 10.

68. Both quotes from John Teresko, "Managing the Process Plan," *Industry Week,* 7 March 1988, p. 35.

69. Maleki, *Flexible Manufacturing Systems,* p. 258.

70. Rohan, "Factories of the Future," p. 33.

71. Ohno, *Toyota Production System,* p. 45.

72. Ibid., p. 47.

73. Ibid., p. 48.

74. Richard B. Chase, "The Service Factory," *Harvard Business Review,* July-August 1989.

75. Hall, "Competitive Manufacturing Infrastructure Issues," p. 18.

## Chapter 7: Shared Dreams

1. John H. Sheridan, "Suppliers: Partners in Prosperity," *Industry Week,* 19 March 1990, p. 14.

2. David N. Burt, "Managing Suppliers Up to Speed," *Harvard Business Review,* July-August 1989, p. 129.

3. Ibid.

4. Sheridan, "Suppliers: Partners in Prosperity," p. 16.

5. Geoffrey Smith, "The Federation Trend," *FW,* 7 August 1990, p. 10.

6. Derek Leebaert, "Later Than We Think," in *Technology 2001,* pp. 10–11.

7. See Nagel and Dove, *21st Century Manufacturing Enterprise Strategy,* vol. 2, p. 12.

8. Author interview with Michael Borrus, January 1992.

9. Michiel R. Leenders and David L. Blenkhorn, *Reverse Marketing* (New York: Free Press, 1988), pp. 7, 23.

10. Thomas M. Rohan, "Supplier-Customer Links Multiplying," *Industry Week,* 17 April 1989, p. 20.

11. Robert E. Spekman, "Strategic Supplier Selection: Understanding Long-Term Buyer Relationships," *Business Horizons,* July-August 1988, p. 77.

12. Interview with Jim Morgan, 3 February 1992.

13. Interview with Michael Borrus, January 1992.

14. Sheridan, "Suppliers: Partners in Prosperity," p. 13.

15. Rohan, "Supplier-Customer Links Multiplying," p. 20.

16. Joel Dreyfuss, "Shaping Up Your Suppliers," *Fortune,* 10 April 1989, p. 116.

17. Ibid.

18. Dell Computer figures per David Webb, "Suppliers Reeling from the Quality Onslaught," *Electronic Business,* 7 October 1991, pp. 107–108; Harris Electronics and Harley-Davidson figures per Kevin Kelly and Otis Port, "Learning from Japan," *Business Week,* 27 January 1992, pp. 52–60; BMW figures per John Templeton, "Grill to Grill with Japan," *Business Week/Quality* 1991, p. 39.

19. Templeton, "Grill to Grill with Japan," p. 39.

20. Hertz and Allegis examples are from Christopher Farrell, "You Buy My Widgets, I'll Buy Your Debt," *Business Week,* 1 August 1988, p. 85.

21. Farrell, "You Buy My Widgets," p. 85.

22. Dreyfuss, "Shaping Up Your Suppliers," p. 116.

23. Stephanie Strom, "U.S. Garment Makers Come Home," *New York Times,* 8 October 1991, p. C1.

24. Cohen and Zysman, *Manufacturing Matters,* p. 148.

25. Dan Marshall, "HP Helps Deliver Just-in-Time," *Harvard Business Review,* July-August 1989, p. 133.

26. E. J. Stefanides, "Turning Suppliers into Partners," *Design News,* 3 July 1989, p. 82.

27. Brian Moskal, "Ah, Togetherness!" *Industry Week,* 18 January 1988, p. 17.

28. Woodruff, "Racy Viper Already a Winner," p. 36.

29. Benn R. Konsynski and F. Warren McFarlan, "Information Partnerships— Shared Data, Shared Scale," *Harvard Business Review,* September-October 1990, p. 117.

30. Templeton, "Grill to Grill with Japan," p. 39.

31. Kelly and Port, "Learning from Japan," p. 59.

32. Sheridan, "Suppliers: Partners in Prosperity," p. 13.

33. Burt, "Managing Suppliers Up to Speed," p. 130.

34. Stefanides, "Turning Suppliers into Partners," p. 82.

35. Larry Armstrong, "Beyond 'May I Help You?'" *Business Week/Quality* 1991, pp. 102–103.

36. Robert Williams, "The Do's and Don'ts of Service," speech given 19–28 June 1990.

37. *Building a Chain of Customers*, p. 34.

38. Womack, Jones, and Roos, *Machine That Changed the World*, pp. 169–193.

39. Robert D. Buzzell and Bradley T. Gale, *The PIMS Principles*, chap. 6 (New York: Free Press, 1987).

40. Keki R. Bhote, *Strategic Supply Management* (New York: AMACOM, 1989), pp. 114–116.

41. Buzzell and Gale, *The PIMS Principles*, p. 103.

42. Barbara Bund Jackson, *Winning and Keeping Industrial Customers* (Lexington, Mass.: Lexington Books, 1985), pp. 14–15.

43. Leonard L. Berry, "Services Marketing Is Different," *Harvard Business Review*, May-June 1980, p. 24.

44. Dana Milbank, "As Stores Scrimp More and Order Less, Suppliers Take on Greater Risks, Costs," *Wall Street Journal*, 10 December 1991, p. B1.

45. Ibid.

## Chapter 8: Rethinking Management

1. John Hillkirk and Gary Jacobsen, "CEO of the Decade," in *Grit, Guts & Genius* (Boston: Houghton Mifflin, 1990), pp. 73–83; and Donald Peterson and John Hillkirk, *A Better Idea: Redefining the Way Americans Work* (Boston: Houghton Mifflin, 1991).

2. Jan Carlzon, *Moments of Truth* (Cambridge, Mass.: Ballinger, 1987), pp. 21–29.

3. Charles D. Wrege and Ronald G. Greenwood, *Frederick W. Taylor* (Homewood, Ill.: Business One/Irwin, 1991), pp. 84–88.

4. Quotes from John Dos Passos, *The Big Money* (New York: Modern Library, 1937), pp. 19–25. Information on Taylor's breakdown from Wrege and Greenwood, *Frederick W. Taylor*, p. 125.

5. Dos Passos, *The Big Money*, p. 55.

6. Ben Hamper, *Rivethead* (New York: Warner Books, 1991), p. 205.

7. Jean-Jacques Servan-Schreiber, *The American Challenge*, trans. Ronald Steel (New York: Atheneum, 1968).

8. Derek Leebaert, ed., *Technology 2001* (Cambridge, Mass.: MIT Press, 1991), p. 25.

9. Dick Cornuelle, *Demanaging America: The Final Revolution*, in Tom Brown, "The De-Management of America," *Industry Week*, 16 July 1990, p. 47.

10. Peter F. Drucker, "Tomorrow's Restless Managers," *Industry Week*, 18 April 1988, p. 25.

11. Al McBride and Scott Brown, "The Future of On-line Technology," in *Technology 2001*, p. 178.

12. Ibid., p. 29.

13. Teri Sprackland, "Middle Managers Seek Escape from Quality Squeeze," *Electronic Business*, 7 October 1991, p. 83.

14. Franklin Mint figures and Westerman estimate per Jeremy Main, "The Winning Organization," *Fortune*, 26 September 1988, p. 52; General Motors figures per Drucker, "Tomorrow's Restless Managers," p. 25; Kodak figures per John S. McClena-

hen, "Flexible Structures to Absorb the Shocks," *Industry Week,* 18 April 1988, p. 41; and Intel figures per Willett, "What Drives Quality at Intel?" p. 28.

15. Stephen Kreider Young, "A 1990 Reorganization at Hewlett-Packard Already Is Paying Off," *Wall Street Journal,* 22 July 1991, p. A1.

16. Ibid., p. A5.

17. Main, "The Winning Organization," p. 50.

18. William Pat Patterson, "Information Systems/Unlimited New Frontiers," *Industry Week,* 7 March 1988, p. 46.

19. Author interviews with T. J. Rodgers. See also Dumaine, "The Bureaucracy Busters," p. 46.

20. Mintzberg, "Thick Management," a review of Henry Mintzberg, *Mintzberg on Management: Inside Our Strange World of Organizations* (New York: Free Press, 1989).

21. Beverly Beckett, Bernie Knill, Thomas M. Rohan, and George Werner, "New Wizards of Management," *Industry Week,* 19 March 1990, p. 62.

22. Robert F. Morison, "The Shape of I/S to Come," *Indications,* July/August 1990, pp. 1–7. In this article Morison uses such terms as "virtual centralization," "virtual staff," and "virtual organization" to describe these activities.

23. From a speech by Debi Coleman, November 1990.

24. Richard O. Mason, James L. McKenney, Duncan G. Copeland, and the Harvard MIS History Project, "Developing an Historical Tradition in MIS Research," draft, 16 May 1991, pp. 28–30.

25. Author interview with Debi Coleman.

26. DEC figures per Beckett et al., "New Wizards of Management," p. 66; HP figures per Patterson, "Information Systems/Unlimited New Frontiers," p. 44; and Security Pacific Bank figures per Rosabeth Moss Kanter, "The New Managerial Work," *Harvard Business Review,* November-December 1989, p. 86.

27. Ian Meiklejohn, "New Forms for a New Age," *Management Today* (UK), May 1989.

28. *Industry Week* report by Therese R. Welter, "IW Readers Play Futurist," *Industry Week,* 20 June 1988, pp. 71–73; Arthur Young study, "Private Study Finds Executive Management Out of Touch with Advanced Technology," *Aviation Week & Space Technology,* 11 April 1988, p. 147.

29. Teresko, "Managing the Process Plan," p. 39.

30. McClenahen, "Flexible Structures Absorb Shocks," p. 42.

31. Teresko, "Managing the Process Plan," p. 39.

32. Beckett et al., "New Wizards of Management," p. 82.

33. Ibid.

34. Shoshana Zuboff, associate professor, Harvard Business School, quoted in Joel Dreyfuss, "Catching the Computer Wave," *Fortune,* 26 September 1988, p. 78.

35. Dumaine, "The Bureaucracy Busters," pp. 46, 49–50.

36. Kim B. Clark, "What Strategy Can Do for Technology," *Harvard Business Review,* November-December 1989, p. 98.

37. Brown, "The De-Management of America," p. 47.

38. Kanter, "The New Managerial Work," pp. 88, 91–92.

39. David Luke III quote per Mark Goldstein, "Just Managing Won't Be Enough," *Industry Week,* 18 April 1988, p. 21; Fred G. Steingraber, "Managing in the 1990s," *Business Horizons,* January-February 1990, p. 61; Ronald H. Walker, "The 21st Century Executive," *Industry Week,* 18 April 1988, p. 12.

40. Quote and Kearney figures per Steingraber, "Managing in the 1990s," p. 58.

41. Walker, "The 21st Century Executive," p. 12.

42. Donald Petersen and John Hillkirk, *A Better Idea: Redefining the Way Americans Work* (Boston: Houghton Mifflin, 1991), serialized in USA Today, September 20, 1991, p. 2B.

43. John W. Gardner, *On Leadership* (New York: Free Press, 1990), p. 24.

## Chapter 9: A New Kind of Worker

1. Information on GM's Oklahoma City and Orion Township plants per Gregory A. Patterson, "Two GM Auto Plants Illustrate Major Role of Workers' Attitudes," *Wall Street Journal*, 29 August 1991, pp. A1–A2.

2. The Lehigh researchers independently came up with the same phrase for this situation, "new social contract." However, as with "virtual company," Nagel and Dove's new social contract model deals with temporary multicorporation link-ups for the design and manufacture of new products—a development we believe will be rarer in the new business environment than do they.

3. Edward Lawler quote per Michael A. Verespej, "People: The Only Sustainable Edge," *Industry Week*, 1 July 1991, p. 19; Lehigh University researchers quote per Nagel and Dove, *21st Century Manufacturing Enterprise Strategy*, vol. 1, p. 10.

4. Nagel and Dove, *21st Century Manufacturing Enterprise Strategy*, vol. 1, p. 10.

5. Bob Davis and Dana Milbank, "If the U.S. Work Ethic Is Fading, Alienation May Be the Main Reason," *Wall Street Journal*, 7 February 1992, pp. A1–A2.

6. Ibid.

7. Robert D. Hof, "Make It Fast—And Make It Right," *Business Week/Quality* 1991, p. 79.

8. Peter Burrows, "Power to the Workers," *Electronic Business*, 7 October 1991, p. 98.

9. William B. Johnston and Arnold H. Packer, *Workforce 2000* (Indianapolis, Ind.: Hudson Institute, 1987), p. xxvii.

10. "View from Top: Lousy Workers," Associated Press, quoted from *San Jose Mercury-News*, 8 April 1991, p. 12A.

11. Unless otherwise noted, all Wiggenhorn quotes are from William Wiggenhorn, "Motorola U: When Training Becomes an Education," *Harvard Business Review*, July-August 1990, pp. 77–78.

12. Johnston and Packer, *Workforce 2000*, data from the Executive Summary.

13. Wiggenhorn "Motorola U," p. 77.

14. Quang Bao and Elizabeth B. Baatz, "How Selectron Finally Got in Touch with Its Workers," *Electronic Business*, 7 October 1991, p. 48.

15. Nancy J. Perry, "The Workers of the Future," *Fortune 1991*, spring issue, p. 71.

16. John Hoerr, "With Job Training, a Little Dab Won't Do Ya," *Business Week*, 24 September 1990, p. 95.

17. Gail E. Schares and John Templeman, "Think Small," *Business Week*, 4 November 1991, p. 62.

18. Ibid.

19. Burrows, "Power to the Workers," p. 100.

20. Thomas J. Murray, "How Motorola Builds in Speed and Quality," *Business Month*, July 1989, p. 37.

21. Wiggenhorn, "Motorola U," p. 80.

22. Michael A. Verespej, "No Empowerment without Education," *Industry Week*, 1 April 1991, p. 28.

23. Perry, "Workers of the Future," p. 71.

**Notes**

. . . . . . . . . . . . . . . . . . . . . . . . . . . . . . . . . . . . . . . . . . . . . . . . . . . . . . . . . . . . . . . . . . . . . . . .

24. Ibid.

25. Quang Bao and Elizabeth B. Baatz, "How Selectron Finally Got in Touch with Its Workers," *Electronic Business,* 7 October 1991, p. 47

26. Willett, "What Drives Quality at Intel?" pp. 38–40.

27. Ibid.

28. Michael A. Verespej, "No Empowerment without Education," *Industry Week,* 1 April 1991, p. 28.

29. Ibid., pp. 28–29.

30. Ibid.

31. Joseph F. McKenna, "Dave Browne's Style: Analysis, but Not Paralysis," *Industry Week,* 3 September 1990, p. 19.

32. The source of this term is business consultant Ed Wilcox.

33. McKenna, "Dave Browne's Style," p. 20.

34. Templeton, "Grill to Grill with Japan," 1991, p. 39.

35. Author interview, 16 October 1991.

36. Perry, "Workers of the Future," p. 69.

37. Johnston and Packer, *Workforce 2000,* p. xxi.

38. Ibid.

39. Barbara Whitaker Shimko, "New Breed Workers Need New Yardsticks," *Business Horizons,* November-December 1990, p. 34.

40. Ibid., p. 35.

41. Ibid., pp. 35–36.

42. H. C. Traiandis, J. M. Feldman, D. E. Weldon, and W. M. Harvey, "Ecosystem Distrust and the Hard-to-Employ," *Journal of Applied Psychology,* February 1975, pp. 44–56. Referenced in Shimko, "New Breed Workers."

43. Perry, "Workers of the Future," p. 72.

44. Shimko, "New Breed Workers," p. 35.

45. Brian Dumaine, "Who Needs a Boss?" *Fortune,* 7 May 1990.

46. Burrows, "Power to the Workers," p. 96.

47. See Dumaine, "Who Needs a Boss," pp. 52–53.

48. Geoffrey Smith, "A Warm Feeling Inside," *Business Week/Quality* 1991, p. 158.

49. Roy L. Harmon and Leroy D. Peterson, *Reinventing the Factory* (New York: Free Press, 1990), pp. 94–95.

50. Harry C. Katz, Thomas A. Kochan, and Jeffrey Keefe, "Industrial Relations and Productivity in the U.S. Automobile Industry," Brookings Papers on Economic Activity, vol. 3, 1988, pp. 685–715.

51. Michael A. Verespej, "When You Put the Team in Charge," *Industry Week,* 3 December 1990, p. 30. The survey was conducted by the Association for Quality and Participation, Development Dimensions International, and *Industry Week.*

52. Michael Schrage, *Shared Minds* (New York: Random House, 1990), p. 29.

53. Tom Brown, "Want to Be a Real Team?" *Industry Week,* 3 June 1991, p. 17. Source of information is Glenn Parker, *Team Players and Teamwork* (New York: Jossey-Bass, 1990).

54. Allan Cox, "The Homework Behind Teamwork," *Industry Week,* 7 January 1991, p. 23.

55. Ibid., p. 21.

56. Verespej, "When You Put the Team in Charge," p. 32.

57. Ibid., p. 31.

58. Ibid., p. 32.

59. Nagel and Dove, *21st Century Manufacturing Enterprise Strategy,* vol. 1, p. 11.

60. Davis and Milbank, "If the U.S Work Ethic Is Fading," p. A2.

61. Author interview with Wilf Corrigan, May 1991.

62. Chandler, *The Visible Hand,* p. 493.

63. Ibid., pp. 205–206.

64. Ibid., p. 493.

65. Ibid., p. 494.

66. U.S. Labor Department, as quoted in "Why Unions Thrive Abroad—But Wither in the U.S.," *Business Week,* 10 September 1990, p. 26.

67. Peter Drucker, *The New Realities* (New York: Harper & Row, 1989), p. 192.

68. Both quotes from Michael A. Verespej, "The Illusion of Cooperation," *Industry Week,* 19 August 1991, pp. 13–14.

69. John Hoerr, "What Should Unions Do?" *Harvard Business Review,* May-June 1991, p. 30.

70. Jane Slaughter, "Is Labor Ready to Break the Siege?" *The Progressive,* December 1989, p. 23.

71. Jonathan Kozol, *Savage Inequalities* (New York: Crown Publishers, 1991), p. 81.

72. Hoerr, "What Should Unions Do?" p. 31.

73. From Lawrence Mishel and Paula B. Voos, "Unions and American Economic Competitiveness," quoted in *Washington Post* Staff, "Don't Blame Unions for Trade Deficit, Book Says," *San Jose Mercury-News,* 15 February 1992, p. 10F.

74. Hoerr, "What Should Unions Do?" p. 39.

75. Lowell Ralph Turner, "The Politics of Work Reorganization: Industrial Relations under Pressure in Contemporary World Markets," doctoral dissertation, University of California at Berkeley, 1991, p. 23.

76. Ibid., p. 182.

77. Ibid., pp. 232–233.

78. Ibid., pp. 397–398.

79. Richard M. Locke, "The Resurgence of the Local Union: Industrial Restructuring and Industrial Relations in Italy," paper presented at BRIE conference, Spring 1991, p. 10.

80. Ibid., p. 23.

81. Turner, "Politics of Work Reorganization," pp. 402–409.

82. Michael A. Verespej, "Labor Will Be the Partner, Not the Enemy," *Industry Week,* 18 April 1988, p. 31.

83. Hoerr, "What Should Unions Do?" p. 39.

84. Republic Engineered Steels story and quotes from Peter Kilborn, "New Paths in Business When Workers Own," *New York Times,* 22 November 1991, pp. A1 and A10.

85. Turner, "Politics of Work Reorganization," pp. 45 and 49.

86. We thank Tim Ferguson of the *Wall Street Journal* for the question.

87. Author interview with Mr. Corrigan.

## Chapter 10: Spreading the Word

1. Joanne Lippman, "Consumers' Favorite Commercials Tend to Feature Lower Prices or Cuddly Kids," *Wall Street Journal,* 2 March 1992, p. B1.

2. See Chandler, *The Visible Hand.*

3. Gretchen Morgenson, "The Trend is Not Their Friend," *Forbes,* 16 September 1991, pp. 115–116.

4. Ibid., pp. 116, 118.

5. Michael J. McCarthy, "Soft Drink Giants Sit Up and Take Notice as Sales of

Store Brands Show More Fizz," *Wall Street Journal,* 6 March 1992, p. B1.

6. Michael de Kare-Silver, "Brand Flakes," *Management Today,* November 1990, p. 119.

7. Liz Levy, "Getting to the Point," *Management Today,* September 1990, p. 106.

8. Regis McKenna, *Relationship Marketing* (Reading, Mass.: Addison-Wesley, 1991), p. 24.

9. Staff interview, "The Great Ad Debate," *Business Month,* September 1989, p. 69.

10. Ibid., p. 69. Figures from Kelly O'Dea, senior vice president of Ogilvy & Mather.

11. Dana Wechsler Linden and Vicki Contavespi, "Media Wars," *Forbes,* 19 August 1991, p. 38.

12. Theodore Leavitt, *The Marketing Imagination* (New York: Free Press, 1986).

13. Levy, "Getting to the Point," p. 106.

14. Author conversation with Dorio.

15. Martin Stein of Abt Associates says that "General Motors' advertising campaign for Mr. Goodwrench, the GM dealer's mythical service expert, doesn't work because people doubt the quality of service being advertised will be available. They may be looking for Mr. Goodwrench, but they aren't finding him." From Davidow and Uttal, *Total Customer Service* (New York: Harper & Row, 1989), p. 81.

16. Ibid., p. 83.

17. Richard Winger and David Edelman, "The Art of Selling to a Segment of One," *Business Month,* January 1990, p. 70.

18. Citicorp, Arbitron, and Nielsen figures and Laurel Cutler information per Martin Mayer, "Scanning the Future," *Forbes,* 15 October 1990, pp. 114–117.

19. Zachary Schiller, "Stalking the New Consumer," *Business Week,* 28 August 1992, p. 62.

20. This list of key factors in the production of good service comes from Davidow and Uttal, *Total Customer Service,* pp. xx–xxii.

21. See also Robert L. Desatnick, *Managing to Keep the Customer* (Boston: Houghton Mifflin, 1987), pp. 15–33.

22. Davidow and Uttal, *Total Customer Service,* p. xxi–xxii.

23. Ibid., p. xxii.

24. Christopher Power, "Getting 'Em While They're Young," *Business Week,* 9 September 1991, p. 94. Figures are from James McNeal, Texas A&M University.

25. Ibid., p. 95.

26. Marian B. Wood and Evelyn Ehrlich, "Segmentation: Five Steps to More Effective Business-to-Business Marketing," *Sales & Marketing Management,* April 1991, p. 59.

27. Ibid., pp. 59–60.

28. Harold J. Novick, "Target Your Sales Channels, Too," *Industry Week,* 6 March 1989, p. 11.

29. Winger and Edelman, "Selling to a Segment of One," p. 79.

30. Ibid.

31. Milind M. Lele, *The Customer Is Key* (New York: John Wiley & Sons, 1987), p. 154.

32. Ibid., p. 96.

33. Ibid., pp. 164–165.

34. Ibid., pp. 173–174.

35. "The Future of ASICS," speech by Wilf Corrigan to ASIC Technology and News, 1 December 1989.

36. Author conversation with Regis McKenna.

37. Christopher Power, "Value Marketing," *Business Week,* 11 November 1991, p. 133.

38. Guy Kawasaki, *Selling the Dream* (New York: HarperCollins, 1991), pp. 3–4.

## Chapter 11: Toward a Revitalized Economy

1. Barbara Kantrowitz, "An F in World Competition," *Newsweek,* 17 February 1992, p. 57.

2. Paul Krugman, *The Age of Diminished Expectations* (Cambridge, Mass.: MIT Press, 1990), p. 1.

3. Sylvia Nasar, "Employment in Service Industry, Engine for Boom of 80's, Falters," *New York Times,* 2 January 1992, p. 1A.

4. Stefan Fatsis (AP), "How Mighty General Motors Fell," *San Francisco Examiner,* 22 December 1991, p. E-7. See also Alex Taylor III, "Can GM Remodel Itself?" *Fortune,* 13 January 1991, p. 27.

5. *Time,* 2 February 1987.

6. From "Management Levers for Successful Globalization," presented by McKinsey & Company to the DataQuest Presidents Summit Conference, Santa Fe, New Mexico, 12 September 1990, pp. 2–25.

7. Nagel and Dove, *21st Century Manufacturing Enterprise Strategy,* vol. 1, November 1991, pp. 4 and 41.

8. Drucker, *The New Realities,* p. 72.

9. John Cunniff (AP), "U.S. Firms Complain of Skills Shortage," *San Jose Mercury-News,* 20 January 1992, p. 11C.

10. Ibid.

11. Theodore R. Sizer, *Horace's School* (Boston: Houghton Mifflin, 1992), p. 32. Also from a conversation with Dr. Sizer, 17 February 1992.

12. Theodore R. Sizer, *Horace's Compromise* (Boston: Houghton Mifflin, rev. ed., 1992), p. 195.

13. "A Conversation with Bob Reich," *Directions* (Thayer School of Engineering/Dartmouth College), Fall 1991, p. 16.

14. Ibid., p. 17.

15. Ibid.

16. Peter Coy, "How Do You Build an Information Highway?" *Business Week,* 16 September 1991, p. 109.

17. Peter Coy, "Why the Highway Won't Reach Home Just Yet," sidebar, *Business Week,* 16 September 1991, p. 112.

18. Ibid.

19. Peter W. Huber, *Liability: The Legal Revolution and Its Consequences* (New York: Basic Books, 1988), p. 4.

20. Peter Brimelow and Leslie Spencer, "The Plaintiff Attorney's Great Honey Rush," *Forbes,* 16 October 1989, p. 197.

21. Nagel and Dove, *21st Century Manufacturing Enterprise Strategy,* vol. 1, from the Foreword.

22. Ibid., p. 4.

# Index

**Index**

. . . . . . . . . . . . . . . . . . . . . . . . . . . . . . . . . . . . . . . . . . . . . . . . . . . . . . . . . . . . . . . . . . . . . . . . . . . . .